RESPONSIBLE GENETICS

Philosophy and Medicine

VOLUME 70

The titles published in this series are listed at the end of this volume

RESPONSIBLE GENETICS

THE MORAL RESPONSIBILITY OF GENETICISTS FOR THE CONSEQUENCES OF HUMAN GENETICS RESEARCH

by

ANDERS NORDGREN

Department of Public Health and Caring Sciences
(Biomedical Ethics),
Uppsala University, Uppsala, Sweden

SPRINGER-SCIENCE+BUSINESS MEDIA, B.V.

A C.I.P. Catalogue record for this book is available from the Library of Congress.

ISBN 978-90-481-5907-9 ISBN 978-94-015-9741-8 (eBook)
DOI 10.1007/978-94-015-9741-8

Printed on acid-free paper

To my father and mother

It is impossible in the modern world for a man of science
to say with any honesty, “My business is to provide knowledge,
and what use is made of the knowledge is not my responsibility.”

Bertrand Russell, 1960.

TABLE OF CONTENTS

ACKNOWLEDGMENTS

This book could not have been written without the support and input from others. Throughout the project I have been supported by Carl-Henric Grenholm, Professor of Ethics, and Ulf G. Pettersson, Professor of Medical Genetics, both at Uppsala University, Sweden. I am very grateful to them.

I am also very grateful to Thomas H. Murray (now President of the Hastings Center, Garrison NY) for accepting me as a Visiting Scholar at the Center for Biomedical Ethics, Case Western Reserve University, Cleveland OH, in the Spring of 1997. He made my stay very worthwhile and opened many doors. During my stay at the Center and afterwards I have also been given great support and good advice by Eric T. Juengst.

Moreover, I would like to thank the following persons for consenting to discuss with me some of the issues in this book at an early stage of my research: George Annas, Adrienne Asch, Jonathan Beckwith, Stephanie Bird, Carl Reinhold Bråkenhielm, Robert Cook-Deegan, Sharon Durfy, Troy Duster, Ezekiel Emanuel, Mark Frankel, Mats G. Hansson, Gunnar von Heijne, Albert Jonsen, Rihito Kimura, Barbara Koenig, Bernard Lo, Eric Meslin, Arno Motulsky, Marie Nordström, Diane Paul, Stephen Post, Danielle Ripich, Mark Sagoff, Gunnar Tibell, Robert Wachbroit, LeRoy Walters, Charles Weiner, and Claes-Göran Westrin.

Let me stress that the views expressed in the book are all mine. They are not necessarily shared by the persons mentioned above.

I am also grateful to the Centre for Research Ethics and the Research Program Ethics in Biomedicine at Uppsala University for giving me the opportunity to carry out the research behind the study in a stimulating academic environment.

The language has been examined by Judith Rinker and I would like to thank her.

Finally, I would like to thank two publishers for allowing me to include parts of papers published elsewhere. Parts of Chapter 1 have appeared in *Theoretical Medicine and Bioethics*. These are published with permission of Kluwer Academic Publishers, Dordrecht (Nordgren, 1998a). Parts of Chapter 6 have been published in *Community Genetics*, and are included with permission of Karger, Basel (Nordgren, 1998b).

Research for the study has been funded by a grant from The Swedish Medical Research Council.

Uppsala, September 2001
Anders Nordgren

INTRODUCTION

MORAL RESPONSIBILITY AND THE HUMAN GENOME PROJECT

The Human Genome Project (HGP) is now almost completed. A draft of the full DNA sequence of the human genome has recently been published (Int. Hum. Gen. Seq. Cons., 2001; Venter et al., 2001). The project started in 1990 and was planned to be completed fifteen years later. It is now clearly ahead of schedule. But even when the project is completed much work remains–the analysis of the function of the 30-40,000 genes in the human genome is beyond the scope of the project and will last for decades.

However, even before the HGP started, it was rather controversial from an ethical point of view; not so much because of the sequencing work as such, but because of its anticipated consequences. The results of the project might be used, for example, in developing genetic tests, genetically tailor-made drugs, and gene therapies. Some people fear that genetic testing may lead to discrimination in insurance and at the workplace. Some view the pharmaceutical companies' research on genetically tailor-made drugs as an unacceptable commercialization of science. Some anticipate that the development of gene therapy will give rise to a society in which parents want to design perfect babies with genetic enhancement techniques. This forces us to ask: should the HGP perhaps never have been initiated in the first place? Or more importantly: how should the implications be handled now when the project is almost completed?

Some of the ethical issues were anticipated at the planning stage of the project. Therefore, right from the start the funding agencies allocated a certain percentage to ELSI (Ethical, Legal and Social Implications) research. Some people view this as a way for the scientific community to take moral responsibility for the HGP and its consequences. Critics maintain that ELSI research is intended merely to gain acceptance of risky technologies, as a way of making the research legitimate.

This dispute raises the more general question of moral responsibility in human genetics research. What does moral responsibility really mean in this kind of research? What moral responsibility lies with the geneticists as individuals and as a community for the choice of topic, experimental work, publication, and application? Responsibility is not just a matter of blame or praise. First and foremost, it is a normative issue: What can reasonably be the moral obligations of scientists doing this kind of research? What moral responsibility lies with the geneticists for its consequences? Do geneticists have a responsibility for trying to affect other people's applications of their findings?

One might think that such questions about moral responsibility have been discussed a great deal. This is, however, not the case. To be sure, the rhetoric of 'scientific responsibility' is prominent in the debate on the research ethics of genetics, but its meaning is very seldom clarified. In the ELSI research, it is commonly the actual and possible consequences as such that have been analyzed and discussed. The issue of the moral responsibility of the geneticists for these consequences has been very much neglected. It is, however, the key issue of this book.

HUMAN GENETICS AND THE ROLE OF HGP

The Human Genome Project has already been and will in the future continue to be of key importance in many different branches of human genetics research including medical genetics, clinical genetics, pharmacogenomics, and gene therapy research. It is an international endeavor coordinated by the Human Genome Organisation (HUGO), founded in 1988. The coordination concerns the laboratory work itself as well as the administration of the rapidly growing amount of information about the genomes of humans and model organisms (McKusick, 1989; Br. Med. Assoc., 1998, pp. 193-194).[1]

The overall objective of the HGP is to map and sequence the human genome. More precisely, the HGP is a project designed to construct genetic and physical maps of the human genome, to determine the complete nucleotide sequence of the human DNA, to localize the genes of the human genome, and to conduct similar analyses on the genomes of some other organisms used as models. Examples of such model organisms are the roundworm C. Elegans, yeast, the bacterium E. coli., the mouse, and the fruit fly (NCHGR, 1991; NHGRI, 1997).[2]

The HGP aims at understanding the structure of the human genes. The understanding of the function of the genes is beyond the scope of the project in the strict sense. But, of course, alongside the mapping and the sequencing, analyses of the function of genes have already started and will continue for a long time after the project is completed (NHGRI, 1997, pp. 61-67). The best way to understand what the HGP is all about is to view it as a kind of tool-making (NCHGR, 1991; Kingsbury, 1997). With the help of the maps and the knowledge of the sequence of the human genome, many other kinds of human genetics research, such as function analysis, genetic drug research, and gene therapy research, can be carried out with precision.

It is important to note that the more specific objectives of human genome research may vary among different countries. In a sense it might be more appropriate to talk about many different human genome projects than a single one.

---

[1] See also URL: www.gene.ucl.ac.uk/hugo/.

[2] See also URL: www.nhgri.nih.gov:80/HGP/ and URL: www.er.doe.gov/production/ober/hug_top.html.

From the point of view of HUGO, however, one may speak about one international project (Cook-Deegan, 1994, pp. 187-230).[3] It is also important to note that the HGP in the narrow sense is publicly funded, or–in some countries–funded by grants from foundations like the Wellcome Trust. However, there are also very important commercial counterparts, not least the sequencing project carried out by Celera Genomics with J. Craig Venter as principal scientist (Venter et al., 2001).[4]

The most extensive mapping and sequencing efforts are made in the US. In much of the literature about the HGP it is the US project that is explicitly referred to, not the international one. The main agencies responsible for planning and developing the project in this country were the Department of Energy (DOE) and the National Institutes of Health (NIH). The planning process led to the publication of a joint research plan in October 1990, signifying the start of the US project (Cook-Deegan, 1994, pp. 161-185).[5]

If we turn to Europe, it is apparent that by the end of the 1980s only a few teams were mapping and sequencing the human genome. Most of the work was general genetics research, including studies of the function of genes and studies of nonhuman genomes. However, in 1990 the research program Human Genome Analysis was formally announced by the European Community Council of Ministers. The effort was concentrated on specific diseases and particularly interesting areas in the human genome (Hallen, Klepsch, 1995). This can also be seen in the BIOMED programs replacing the original Human Genome Analysis program (Hallen, 1998).

In the literature on the HGP there are numerous references to 'the' human genome. This signifies the 'basic' human genome, i.e., the standard genetic information contained in a typical composite of the 'normal' human chromosomes. This 'basic' human genome is an abstraction that can be compared to the human skeleton. However, it should be noted that there is no ideal example of the genome. In fact, much of the interest in human genome research has to do with variations among individual humans. The HGP is often justified by reference to the identification of disease genes, i.e., genes that deviate from the 'standard' genome.

There are other kinds of human genome research beside the HGP. One is function analysis, which has already started but will continue for many decades after the project in the strict sense is completed. Another kind is 'population genomics', for instance within the Human Genome Diversity Project (HGDP). The aim of this kind of research is to collect and analyze genetic information from ethnic groups around the world.[6]

---

[3] See also URL: www.gene.ucl.ac.uk/hugo/.

[4] See also URL: www.celera.com, and URL: www.incyte.com.

[5] See also URL: www.nhgri.nih.gov:80/HGP/ and URL: www.er.doe.gov/production/ober/hug_top.html.

[6] See URL: www.stanford.edu/group/morrinst/hgdp/summary93.html. See also British Medical Association, 1998, pp. 195.

The tools made available by the HGP, however, are used not only in function analysis and population genomics, but also in clinical genetics, pharmacogenomics and gene therapy research. Such research was, of course, to a varying extent carried out even before any useful results emerged from the Human Genome Project. It is obvious, however, that the genetic knowledge provided by the HGP gives it much better tools. It is therefore reasonable to view such research as applications of the knowledge gained through the project. Another reason for this is that in applications for funding of human genome research these kinds of applications are often referred to as expected beneficial outcomes of the research.

ETHICAL IMPLICATIONS

Human genetics research has important ethical implications for many different agents and parties such as individual patients and their families, different social and ethnic groups, doctors and genetic counselors, public and private health care, biotechnological companies, insurance companies, employers, the government, the general public, and, of course, the geneticists themselves. In this study, I shall focus primarily on the moral responsibility of the geneticists. The reason for this is, as I have stressed above, that this aspect is very seldom reflected on in the ELSI literature. The possible use and abuse of the applications in society and health care have been discussed much more thoroughly.

The Moral Responsibility of the Geneticists

What responsibility do the geneticists have for their choice of subject (topic), the research process in the narrow sense (empirical and theoretical work), the publication of the results, and the application of these results? As the subtitle of the book indicates, I focus in particular on the responsibility of geneticists for the consequences of their research. With regard to these consequences, it is sometimes useful to make a distinction between those that are direct and those that are indirect. The distinction, however, is not strict. The direct consequences of research are those appearing in the research process in itself, for example, the consequences for human subjects participating in a study. The indirect consequences are those of the results of the research after publication, for example, the application of genetic knowledge in other kinds of research or in society. However, what counts as direct and indirect consequences depends on exactly what kind of genetics research we are talking about. For instance, the consequences for human subjects participating in gene therapy research are direct consequences of this research, while the gene therapy research itself, at least to some extent, can be viewed as an indirect consequence of genome research, since it uses knowledge gained in genome research.

Moreover, it is sometimes essential to make a distinction between the responsibility of geneticists as individuals and as a community, between the responsibility of geneticists in the beginning of their scientific career and those high in the academic hierarchy, and between the responsibility of geneticists working at universities and those working in biotechnological companies. One very important aspect concerns professional policy. What policy should the genetics community have on, for instance, the treatment of human subjects or the applications of genetic knowledge in society? It should be noted that the professional policy of the genetics community is not necessarily identical to the public policy of the government.

This study is in two respects wider in scope than is common in the ELSI literature. As a background to my discussion of the ethical implications of human genetics research, I will devote the first two chapters to more general issues. In Chapter One, I shall discuss the concept of moral responsibility in general. I shall also propose a new ethical framework based on certain empirical findings of the relatively new scientific field called cognitive semantics. Cognitive semantics stresses the metaphorical character of key human concepts. This ethical framework sheds light on the discussion of the ethical implications of human genetics research. It helps us to understand what ethical dispute is all about in this context, and provides a starting point for normative suggestions. In the second chapter, I shall discuss research ethics. Here I shall focus on the concept of moral responsibility of scientists in general. This will put the moral responsibility of the geneticists into a broader perspective. What do we mean by 'moral responsibility' in this specific context? For what are geneticists responsible? To whom are they responsible? Are there any differences in the moral responsibility of scientists doing human genetics research compared to the responsibility of other scientists?

In discussing the moral responsibility of geneticists, it is important to make a distinction between internal and external perspectives. Internal perspectives are the perspectives of the geneticists themselves. External perspectives are those of other people such as philosophers, bioethicists, sociologists, journalists, politicians, the general public, etc. Both kinds of perspectives are legitimate. The internal perspectives are legitimate because it is the geneticists who are carrying out the research. Certainly, they are entitled to make moral judgments on their own work. The external perspectives are legitimate because human genetics research is a social activity influencing the remainder of society in various ways, because it uses the societal infrastructure, and because it is to a large extent funded by taxes. Both kinds of perspectives are also necessary in order to work out well-considered ethical guidelines and policies. Each kind has its advantages as well as disadvantages. The geneticists run the risk of becoming 'home blind' and only seeing their own side of things. People with external perspectives might have a broader societal perspective. A disadvantage of the external perspectives, however, is limited knowledge of scientific facts and limited experience of scientific practice. Deep knowledge and first-hand experience is, on the other hand, the advantage of the internal perspectives. Dialogue is the best way of handling this situation. In this study, I shall discuss the moral responsibility of scientists conducting human genetics research but

I will do this from an external perspective since I am a bioethicist, not a geneticist. The book should be viewed as a contribution to such a dialogue.

What is more precisely the role of bioethicists in this dialogue? Some would argue that their only task is to analyze the ethical problems and the arguments. Others maintain that the bioethicists should also put forward normative proposals. With some hesitation, I support the latter view. Bioethicists have a special training in analyzing ethical arguments as well as knowledge of ethical theory and may therefore to some extent contribute to clarifying the discussion on the ethical implications of genetics. This training and knowledge may also be of some help in constructing normative proposals. However, this does not guarantee that the normative proposals of the bioethicists are more well-founded than other peoples', or that their views are more interesting. In this book I have an analytical as well as a constructive approach. I will analyze arguments related to many different ethical issues raised by genetics but I will also put forward certain normative proposals.

It is important that the bioethicists do not carry out their analyses in the abstract without contact with what is really going on in science. For instance, they should be aware of the struggle for power that affects the development of research. Genetics research is not only a value-laden activity, but also an activity situated in a 'power context' as are all other social activities. This can, for example, be seen very clearly in the book *The Gene Wars* by Robert Cook-Deegan, which is the most authoritative study to date on the origins of the HGP in the US. Already the metaphor in the book's title illustrates the point (Cook-Deegan, 1994).

Why Focus on Moral Responsibility?

Throughout this study I focus on the concept of moral responsibility. There are several reasons for this. One is that the concept of responsibility appears to be particularly relevant to problems in applied ethics, because it has such a central social dimension. Applied ethics does not merely concern the individual who acts. Our acts affect other people and they react in their turn. Acts are performed in a social context. We are always responsible 'to' other parties for our actions. Moreover, we often act together with other people and belong to more or less clearly defined social groups. This raises the question of a shared responsibility. All this is also true of scientists. Scientific research is a social activity and arguably the members of the scientific community have a shared responsibility for its activities.

Another reason for focusing on the concept of responsibility are the immensely increased powers of humans to affect their own future as well as the future of the planet at large. These powers are to a large extent due to technological advances made on the basis of scientific research. One clear example is human genetics and its applications, the main topic of this book. This increased technological power brings the issue of responsibility to the forefront of the ethical discussion, as, for instance, the philosopher Hans Jonas has argued (Jonas, 1984).

A third reason is that the concept of responsibility seems especially well suited to a pluralistic society. It allows for the large gray zones so typical of a society with many different and often conflicting values and norms. The concept of responsibility appears particularly useful in the discussion of the ethical implications of human genetics research. This research continuously raises new ethical problems and there is often no consensus regarding their solution. Different parties might agree that they should act responsibly, although they might have somewhat divergent views on what this means more precisely. The notion of moral responsibility, however, may be a possible meeting-place for people with different value commitments and perhaps even constitute a common ethical platform.

However, 'responsibility' is an elusive term. It is often used, but seldom defined. An important task in this study will therefore be to analyze its meaning. I shall devote some space in Chapter 1 to a general analysis, and some space in Chapter 2 to an analysis of the concept in a scientific context.

The title of this book, *Responsible Genetics*, illustrates the centrality of the concept of responsibility for my discussion of the ethical implications of human genetics research.[7] This might lead the reader to think that the book primarily proposes certain restrictions on genetics research and its application. Let me stress, however, that although one important task for bioethics certainly is to suggest certain restrictions, responsibility is not just to say 'no'. It can be to say 'yes' as well. The title of the book is intended to include even such a positive view.

Policy Declarations and Conventions

Some of the most important ethical questions raised by human genetics concern the professional policy of the genetics community. Ethical considerations are at the very basis of policy-making. In my discussion of ethical issues in human genetics, I will, therefore, now and then discuss certain policy declarations and conventions. Examples are the *Convention on Human Rights and Biomedicine* issued by the Council of Europe, UNESCO's *Universal Declaration on the Human Genome and Human Rights*, and CIOMS's *Declaration of Inuyama*.[8] The views of these ethical codes diverge to some extent, and, of course, their normative statuses differ. Both these aspects, however, make them interesting in an analysis of the issue of moral responsibility in human genetics.

In this study, I will argue for a kind of casuistry, i.e. a case-by-case approach to ethics. But does not such an approach exclude policy? I do not think so (cf. Strong, 1997, p. 75). Sensitivity for the particularities of individual cases does not rule out a recognition of common characteristics. There are not only individual cases, but also

---

[7] Note, however, that the title does not allude to the "Council for Responsible Genetics". Cf. www.genewatch.org.

[8] See Council of Europe, 1996, URL: www.unesco.org/ethics/, and CIOMS (Council for International Organisations of Medical Sciences), 1995.

types of cases. Moreover, ethical principles can be viewed as rules of thumb regarding prototype cases or types of prototype cases. They do not have to be interpreted as universal and absolute. It is quite possible for policy documents to express a need for a context-sensitive balancing of conflicting principles. Furthermore, casuistry in biomedicine does not necessarily have to focus exclusively on clinical issues related to individual patients in special circumstances. It is completely possible from a casuistic point of view to address issues of broad social concern as is commonly done in policy documents. As Baruch Brody has recently shown, a more or less casuistic approach can actually be found in many official policy documents on biomedical research (Brody, 1998, pp. 202-208). Policy excludes a casuistic approach only if policy implies universal and absolute ethical principles, and this not at all necessary.

In discussing policy issues related to genetics, I shall primarily focus on the professional policy of the genetics community, not on the public policy of the government. In doing so, I will, however, indirectly discuss public policy. An important question concerns the direction in which the genetics community should try to affect public policy. The reason for focusing on professional instead of public policy is that this study as a whole focuses on the responsibility of the geneticists, not that of politicians.

Aims of the Study

In short, the objectives of this study are:
(1) to analyze the concept of moral responsibility in human genetics research,
(2) to pinpoint ethical issues that responsible geneticists should take into consideration, and
(3) to present normative suggestions with regard to some of these issues, as a contribution to an interdisciplinary dialogue.
In order to reach these goals, I will analyze fundamental bioethical concepts and principles on the basis of the empirical findings of cognitive semantics and propose a new ethical framework, 'imaginative casuistry', taking these empirical findings as a point of departure.

Let me also very briefly answer two possible objections to the general approach of the study. The first concerns scope. Some critics might say that I want to do too much, and that I should limit myself to just a few ethical problems raised by human genetics and examine them more deeply. However, there are already quite a number of books and articles written with such more restricted goals, and there is certainly a need for a book with a more general and integrative approach. Some ethical issues are best approached broadly. This is clearly the case with the issue of the moral responsibility of geneticists.

The second objection concerns method. Some critics might say that the analytic method used is too imprecise to fulfil the demands of academic philosophy.

However, let me point out that the study is not intended to be a piece of philosophy in a narrow academic sense. The objective is not to present a deep, technical analysis of philosophical problems raised by human genetics. On the contrary, the study should be understood as a contribution to an interdisciplinary dialogue. This means that too technical a language must be avoided.

Outline

With all this in mind, the first task in this study will be to analyze the concept of moral responsibility. This will be done in Chapter 1, where I shall also propose a new ethical framework. The second task will be to discuss in general terms the issue of moral responsibility in science (Chapter 2). After that, the remainder of the book is devoted to different ethical considerations which scientists doing human genetics research should reasonably take into account.

In Chapter 3, I shall discuss various aspects of the Human Genome Project that responsible geneticists should consider. First, I discuss the choice of subject as such. Can the HGP be justified, and if so, how? I also discuss the methods of promotion of the project, not least the use of metaphors. Finally, I turn to the issue of access to the results of the project. A central question in this section concerns gene patenting.

Chapters 4 and 5 focus on certain direct and indirect consequences of various types of human genetics research which responsible geneticists should take into account. How should human subjects and their families be treated in genetic studies? How should small ethnic groups be treated in order to justify the Human Genome Diversity Project being carried out? What ethical problems arise in the application of genetic knowledge in genetic testing and screening? Are there any special ethical problems raised by research on genetically tailor-made drugs? What ethical problems are raised by somatic gene therapy research? Would it be morally responsible to carry out germline gene therapy research and genetic enhancement research? I shall also discuss the use of animals in human genetics research, for example as disease models.

The final chapter (Chapter 6) concerns the application of genetics in reproductive medicine ('reprogenetic medicine'). What should be the long term goals of reprogenetic medicine? What should the professional policy of the genetics community take into account? An important question in this context concerns eugenics. For many geneticists in the early decades of this century, social responsibility implied that eugenics should be promoted. Today most geneticists would argue that social responsibility requires that eugenics be counteracted. This means that, for historical reasons at least, the issue of eugenics is relevant to the issue of the moral responsibility of human geneticists.

CHAPTER ONE

MORAL RESPONSIBILITY AND MORAL IMAGINATION

1. THE METAPHOR OF MORAL RESPONSIBILITY

The rhetoric of 'responsibility' is prominent in the research ethics debate, not least in discussions about genetics and genetic engineering. It is also very common in political and religious discourse. But the term is elusive. It is seldom defined. It is also a rather new term in Western ethical discourse. The term 'responsibility' first appeared in English, German and French in the 17th century (Jonsen, 1968, p. 3). However, it has its etymological roots in the Latin word 'respondeo', which actually has two different but connected root meanings. In the first sense, the term has to do with 'answering'. For example, it could be a matter of giving an answer when called to appear in a legal court or another tribunal. In the second sense, the word 'respondeo' is related to 'accounting'. We owe things to others due to promises or financial debts. It is also possible to see accountability as a special form of answering to others (Schweiker, 1995, pp. 55-56).

In the moral context, 'responsibility' is used metaphorically. It is not a matter of actually answering in a legal court or actually being in a state of financial debt. It is only metaphorically so. The term is used metaphorically in ways corresponding to each of the two root meanings. When the focus is on 'answering', we have an aural/oral metaphor. It is a matter of listening to others or to one's own self or conscience and providing an answer. A typical question is: why did you do it? (cf. Lucas, 1993, pp. 5,6). Another question could be an implicit one in the form of a need of another person that demands some kind of response. Let us call this the 'response' metaphor. When the focus is on 'accounting', we have a financial metaphor. We owe things to others and others owe things to us. We can be in a state of moral obligation to act in a particular way. Let us call this the 'owing' metaphor. In practical discourse, these two metaphors of responsibility are often mixed.

A Variety of Uses

The word 'responsibility' is used in many different ways. The most basic use is perhaps 'being responsible'. A person is responsible for an act or an act of omission.

In a similar way, we talk about 'bearing responsibility' and 'having responsibility'. Sometimes we refer to causal responsibility. A person is causally responsible if he or she has caused an action by free will. At other times we refer to moral or legal responsibility. When we use the phrase 'being responsible' as well as most of the other phrases below, we often add 'to' and 'for'. A person is responsible *to* someone *for* something.

We also talk about 'holding someone responsible'. Often 'to hold someone responsible' is to blame or praise him or her because of something the person has done. A typical example is when we blame someone for the harm resulting from that person's actions. This assigning of responsibility is backward-looking. However, the assigning of responsibility may also be forward-looking. We may say to someone: "In the future, we will hold you responsible whatever happens."

Another use of the term 'responsibility' has to do with 'taking responsibility'. It is a matter of accepting or 'assuming' responsibility. Even here we find both backward-looking and forward-looking uses of the phrase. We may take responsibility for something that has already happened, but we may also take responsibility for what will happen in the future.

Sometimes we talk about 'having responsibility'. A person may have responsibility for an activity or an institution. In such cases, the person has often been appointed to a special social position or social role.

Often, responsibility is just shorthand for 'obligation'. 'To have a responsibility' is to have an obligation. This obligation can be a matter of obeying strictly defined rules, or can refer to a responsibility for a certain area or domain where it is not predetermined in detail how one should decide or act in each particular situation.

The word is also used in a normative sense. It is a matter of 'being a responsible person' or 'acting in a responsible way'. In the former case, responsibility is a virtue. A responsible person is someone with certain good traits of character, and who acts in a particular way. Such a person is often considered to be 'responsive' to the needs of others. When we talk about 'acting in a responsible way', 'responsible' refers to the formal characteristics of good behavior. The expression may also be used as an imperative–'Be responsible!'–meaning that we should take our actions seriously and act on the basis of conscious deliberation.

Below I shall discuss each one of these uses more extensively. Let me first, however, make a comment on the distinctions among responsibility as a matter of obligation, as a matter of virtue, and as a matter of blame and praise.

Some might think that what is important in research ethics are the aspects of obligation and virtue, not that of blame and praise. They would argue that what is interesting is what geneticists should do or not do from an ethical point of view, not whether geneticists are to be blamed or praised for what they are doing. I agree to some extent but not entirely. The moral obligations and virtues of geneticists are of central importance but the issue of blame and praise is also vital. Scientific research is a social activity. Some social groups blame scientists for some aspects of what they are doing, other groups praise them. This is obvious not least with regard to 'hot' topics such as genetics and animal experimentation. We need an understanding of what this means more precisely. A key issue is: is there a 'real' moral responsibility beyond this socially attributed one, or is the socially attributed moral responsibility all there is?

Being Morally Responsible

'Being morally responsible' (or 'bearing moral responsibility' or 'having moral responsibility') is no doubt the aspect of responsibility that has received the most attention in the philosophical debate. In a sense it is also the most basic one. One's view on this aspect will to a large extent determine one's view on the other aspects. For these reasons, I will devote more space to this aspect than to the others.

The philosophical debate on moral responsibility has almost completely focused on the issue of free will and determinism. This can be seen, for instance, in the *Cambridge Dictionary of Philosophy* where the term 'responsibility' does not have a place of its own but is discussed under the heading of 'Free will problem' (Audi (gen. ed.), 1995). However, this approach is not particularly fruitful in a discussion of moral responsibility in science. An approach that brings social aspects to the discussion is preferable.

Agential Models and Social Models

We may distinguish two main kinds of models of moral responsibility, both of which address the question of the ground of moral responsibility. The first kind of model focuses on causal agency. Individual moral responsibility for a particular state of affairs, for example harm to other people, depends on whether the individual has caused it by an act or an act of omission. Causal agency determines in itself moral responsibility. I will call these kinds of models 'agential models' (cf. Schweiker, 1995, p. 78).

According to agential models, causal agency and moral responsibility constitute one single moral fact. If an individual agent has caused the state of affairs, then he or she is morally responsible for it. This means, in addition, that moral responsibility should be kept separate from social blame and praise (see e.g. Feinberg, 1970, pp. 30-31; Frankena, 1973, pp. 71-72; Glover, 1970, p. 96). Moral responsibility is 'real' responsibility in distinction to merely socially ascribed responsibility. Moral blame or praise on the one hand, and social blame or praise on the other, might differ. Moral responsibility is a moral fact to be discovered, although established social practices may make it difficult to do so.

Ontologically, agential models of moral responsibility are based on an individualist account of the relation between individuals and society. They presuppose that the individual self in a fundamental sense is independent of society ('social atomism'). Since the individual is independent, the only morally relevant aspect is whether he or she has caused the state of affairs. Whether society ascribes or does not ascribe responsibility is simply irrelevant from a moral point of view.

The second kind of model focuses on social role. Here the ground of moral responsibility is not causal agency in itself but social practices of blame and praise. This means that moral responsibility is not separated from socially ascribed responsibility (see e.g. French, 1992, pp. 55-70; Smiley, 1992). Let us call these models 'social models' (cf. Schweiker, 1995, p. 86). They give a picture of responsibility that differs substantially from that of agential models. We can rightly hold individuals fully responsible for a state of affairs that they did not directly or even indirectly cause by an act or an act of omission. This is because social practices of blame and praise are linked to social roles. Moral responsibility is a social construction, not a simple moral fact. It is ascribed by the community or a particular social group, not discovered. The content of responsibility is what one's social position or role demands. If one does not live up to this standard, one may be blamed.

In distinction to agential models, social models of moral responsibility are not based on an individualist ontology. On the contrary, they presuppose that the individual self is socially constituted ('social holism'), the product of social influences. Without a community there is no individual self. And because this is the case, moral responsibility is socially determined. Causal agency does not in itself determine moral responsibility. Responsibility depends on social role.

The difference between the two kinds of models can be illustrated by the example of a junior scientist who presents poorly substantiated experimental results. Who is morally responsible, the junior scientist or the leader of the research team?

From the perspective of agential models, it is natural to take the distinction between direct and indirect causal responsibility as a starting point, and state that the junior scientist is directly causally responsible because he is the one who has actually carried out the experiments, while the research leader is only indirectly causally responsible because he gave improper or insufficient instructions to the junior scientist. Given these facts of causal responsibility, the junior scientist is fully morally responsible while the research leader is not.

From the point of view of social models, on the other hand, it is possible to maintain that the research leader can be fully morally responsible although he is only indirectly causally responsible. It is also possible to state that the junior scientist is not fully morally responsible in spite of the fact that he is directly causally responsible. What is decisive are the social roles of the research leader and the junior scientist, not the causal relations. The social role of the research leader implies that he is fully morally responsible for whatever happens in the work of his team, while the junior scientist has no such social role.

A Modified Social Model

Both kinds of models, as they have been presented here, neglect the metaphorical character of the concept of moral responsibility. However, they can both be interpreted as focusing on two different aspects of the 'being responsible' metaphor, namely causal agency and social role. Moreover, both aspects have their valid points. Both represent important aspects of moral responsibility. The problem is how to combine them. In my opinion, social role is the primary in the sense that the moral relevance of causal agency should be determined within the framework of social role (practice). Causal agency is commonly relevant to moral responsibility but does not in itself determine it. My proposal is therefore best described as a modified social model. It avoids the weaknesses of both kinds of models, while acknowledging their valid aspects.

What are the weak aspects of the agential models of moral responsibility? First, the agential models do not recognize the social character of moral responsibility. This is seen in their separation of 'real' moral responsibility from socially ascribed responsibility. There is no completely compelling argument against this separation but there are at least some good reasons. The first is that the question of whether there is a 'real' moral responsibility is theoretically open. There are many conflicting views on the matter. There is no general consensus, and one view might be as good as another. Another reason is that the question is practically meaningless. There is no intersubjective method to settle the matter in practice. The question is also practically uninteresting. What is really important in this world is what beliefs social groups actually have and try to implement in practice. Combined, these reasons suggest that moral responsibility has no backing beyond the community.

Moreover, agential models conflate causal agency and moral responsibility. Causal agency does not in itself determine moral responsibility. It is obvious that individuals, due to their social role, can be morally responsible even when the causal connection to a state of affairs is very indirect or even nonexistent. And conversely, individuals might be only to a very limited extent morally responsible even when they have very directly causally contributed. There are commonly certain causal connections, but a causal connection does not in itself determine moral responsibility. Causal agency should be understood within the framework of social practice (role).

The agential models also wrongly assume that moral responsibility is a matter of discovery. On the contrary, moral responsibility is a matter of ascription (and, as we shall see, assumption). Such ascription of responsibility is based on causal as well as value judgments. In the causal judgment one makes an interpretation of a situation and focuses on a particular part of a causal chain or a particular link in a causal network instead of other parts or links. In doing so, a particular individual may be singled out as *the* cause of a state of affairs. The individual's causal contribution to the state of affairs may indeed be very indirect. This is so because in framing the situation, different individuals are given different social roles, and these roles may determine the ascription of causal responsibility. In the value judgment, moral blame or praise is given to the individual who has been selected as causal agent. He or she is proclaimed to be morally responsible, not only causally responsible.

In addition, it should be noted that in the ascription of moral responsibility, there are some clear cases on which all agree, but there are also complex, difficult cases where different social groups may make different judgments.[1] Differences in judgment may depend on different interests or a struggle for power. This is not recognized by the agential models, either, but it is very important.

Agential models neglect not only that moral responsibility is a matter of ascription. They also neglect the fact that moral responsibility may be a matter of assumption. An individual agent may take responsibility for a state of affairs that he or she has not caused and for which he or she is not ascribed responsibility by the community.

Finally, the ontological view of agential models is unsatisfying. The individual self is not an entity existing independently of society. It is socially and biologically constituted and situated–an individual agent is always a part of a social network and a biological context. There is an interplay between social and biological factors in shaping the individual.

This brings us to the weak aspects of the social models as they are commonly understood. One weak point is that social models, while rightly stressing that the individual self is socially constituted, often do not recognize the influence of biological factors. This influence has been stressed by sociobiology (Wilson, 1978) and evolutionary psychology (Barkow, Cosmides, Tooby, 1992). One may question particular details in their account of how evolution and genes affect individuals and societies. One may also question the sometimes very simplifying view of the interplay between genes and environment. But there is no doubt that biology plays a vital role in the process of shaping individual agents (Bateson, Martin, 2000; Nordgren, 1994).

Moreover, social models, particularly those that form a part of a communitarian political view, often do not fully acknowledge that the individual is a unique combination of different social–and biological–influences. This uniqueness is important because it makes it possible for the individual to criticize socially established practices of blame and praise ('modified social holism'). An example could be the work for the elimination of social roles and traditions that discriminate (May, 1992, pp. 176-183). The notion of a uniquely constituted individual self also makes a substantive concept of individual autonomy possible that is not atomistic. We may talk about individual autonomy in a sense that recognizes a social network. This is important, for example, in the discussion of the reproductive autonomy of prospective parents, a 'hot' issue in clinical genetics.

---

[1] For the distinction between prototypical and nonprototypical cases, see pp. 21-22.

Social models rightly stress the social ascription of responsibility but they often neglect the possibility of individual assumption of responsibility. There is a possibility for individuals to assume a responsibility that deviates from the prevailing one in their communities and social groups. This is an important condition for social criticism.

Finally, it is not recognized by extreme proponents of social models that there are factual constraints on the ascription of causal responsibility. The aspect of social construction is overinterpreted. We cannot simply ascribe causal responsibility to anybody; there must be at least some factual, causal connections between the individual and the state of affairs. Reality puts certain constraints on the causal ascription. But these constraints can only be found in a pragmatic way. Note, however, that I am talking here about the ascription of causal, not moral, responsibility. Moral responsibility can be ascribed without reference to causal responsibility as part of a social role.

In sum, the proposed modified social model of moral responsibility has the following characteristics:

(1) Causal agency is commonly relevant to moral responsibility but does not in itself determine it.

(2) Moral responsibility is ascribed or assumed by communities, social groups, or individuals.

(3) Individuals can ascribe or assume a moral responsibility that deviates from established social practice.

Responsibility 'of'

This brings us to another important aspect of responsibility, namely responsibility 'of'. Commonly responsibility is discussed in individualist terms–it is an individual who is responsible. However, within the modified social model of responsibility outlined here, it is also meaningful to talk about a 'shared' responsibility. Individuals often cooperate on different projects, something particularly obvious in scientific research. Scientists have not only an individual responsibility but also a shared responsibility as members of research groups and the scientific community at large.

Actually, most of our responsibilities are shared rather than uniquely our own (May, 1992, pp. 37-40). But within the shared responsibility, the responsibility of individual members of a group may vary depending on the individual's place in the group. Experienced members and junior members might have different degrees of responsibility. This kind of shared responsibility is distributed, but we may also talk about nondistributed shared responsibility where each member has the same responsibility. In this case we may talk about 'collective' responsibility, although I prefer not to do so for the reason that ultimately only individuals, not collective entities, can have moral responsibility. Organizations do have legal responsibility, but hardly moral responsibility. When we say that a group or organization has moral responsibility, we mean that its members have a shared responsibility.

Moreover, as Larry May has argued, this shared responsibility not only concerns actions by the group but also the attitudes and the 'climate' within a group. By accepting the attitudes and the climate we share responsibility for them (May, 1992, p. 47).

Responsibility 'to' and Responsibility 'for'

When we use the phrase 'being responsible' as well as most of the other phrases described below, we often add 'to' and 'for'. Gabriel Moran has found an interesting historical shift in the use of these expressions. Before the 19th century the emphasis was on 'responsible to', while from the 19th century onwards the emphasis is on 'responsible for' (Moran, 1996, pp. 33-56). However, as far as I understand, there are hardly any clear distinctions between the uses in everyday language. At least today, the terms are often used very laxly. Moran, however, stresses the importance of making a clear distinction between 'responsible to' and 'responsible for'. Therefore, he proposes 'a grammar of responsibility', the title of one of his books, in order to maintain such a distinction. He suggests that the expression 'responsible to' should be used with reference to, for example, individuals, corporations, animals, and even the environment. 'Responsible for', on the other hand, should only be used with reference to acts and acts of omission, not to people, society, the environment, and future generations, which is often the case today (Moran, 1996, pp. 70-78, 109-198).

Moran is too purist. In most cases, the somewhat inconsistent everyday language is quite acceptable. However, I find his proposal useful for my intentions in this book. Therefore, I will use his distinction between 'responsible to' and 'responsible for' in the analytical model of the moral responsibility in science, which I will develop in the next chapter.

Holding Someone Responsible

The different models of responsibility have direct implications for the issue of 'holding someone responsible'. According to social models, being responsible is always a matter of being held responsible by other persons (or having assumed responsibility). Responsibility is always ascribed to someone by other people (or assumed by the agent himself or herself). According to agential models, on the other hand, a person can be responsible without being held responsible (or having assumed responsibility). He or she is morally responsible due to the causal agency in itself. This means that a person can be rightly or wrongly held responsible in an objective sense. There is a 'real' moral responsibility that might diverge from the ascribed one. But according to social models, responsibility is always related to social groups and their practices of blaming and praising. We should not talk about a 'real' responsibility in distinction to an ascribed responsibility. In the proposed modified social model, responsibility is always a matter of ascribing or taking responsibility.

The ascription of responsibility is often backward-looking. We blame or praise someone because of something he or she has done. However, the ascription of responsibility may also be forward-looking: "In the future we will hold you responsible whatever happens." Such forward-looking ascription is often linked to a particular social role. For instance, as a research leader you are expected to do certain things and you are responsible for this.

People are commonly held responsible for actions. But what about omissions? This question is of great relevance to the issue of moral responsibility in science. In order to answer it, we have to take a closer look at the concept of omission. I discern at least two different meanings:
(1) to intentionally not perform an act one should perform, and
(2) to not perform an act one should perform.
When we use 'omission' in the first sense, a person's intention is a cause of the omission. In the choice between performing and not performing an act, one chooses not to perform it. There is a causal connection between intention and omission: the intention causes the omission. This also holds true for the consequences of the omission, to the extent they are intended.

When we use the term in the second sense, the person unintentionally neglects to act. In this case, there is no causal connection between intention and omission because there is no intention not to act. This is also true concerning the consequences of these omissions.

As far as I can see, it is only possible to talk about causal responsibility regarding *intended* omissions, i.e. omissions in the first sense, and their *intended* consequences. There is a causal connection between intention and omission. Regarding omission in the second sense, there is no causal connection and therefore we cannot speak about a causal responsibility.

What about the ascription of moral responsibility? People can clearly be held morally responsible for omissions in the first sense, i.e. intended omissions. An example is an explosion in a lab that could have been prevented if a technician had fulfilled his duty to check the equipment. He was aware of his duty but ignored it. By his omission, the technician was a contributing cause of the explosion. Whether he actually should be held morally responsible depends on the reasons for his neglect. Mere convenience or lack of time would not be sufficient to free him from moral responsibility, but helping a badly injured person to the hospital might.

The question whether one can be held responsible for omissions in the second sense, i.e. for not performing an act one should perform, is more difficult. Can a person truly be held morally responsible for not performing an act he should perform even if he has no intention not to perform it? It seems that moral responsibility may sometimes be ascribed even if there is no causal responsibility in this sense. This is the case when a person is not aware of a duty to perform an act but should have been. This presupposes that he could have been aware of it, i.e. that it is within the scope of his ability. 'Should' implies 'can'. Thus, regarding omission in the second sense one cannot talk about causal responsibility but certainly about moral responsibility. Take, for example, the technician mentioned above. If he was not aware of his duty to check the equipment at the lab–perhaps because he forgot to read the duty list–but should have been aware of it, then he can be held morally responsible for the explosion.

Can we be held equally morally responsible for acts and omissions? Strict consequentialists would say that there is no morally relevant difference. What is decisive is what actually happens, not whether what happens is the result of an act or an omission, or whether what happens is intended or not. Others would say that there is a morally relevant difference, that there are degrees of moral responsibility. I agree. We have a greater moral responsibility for acts than omissions. There is a certain asymmetry. This implies that a distinction should be maintained between acts and omissions.

My main argument stresses the 'homeostatic' character of acts. This character has in particular been pointed out by John R. Lucas. He argues that an unlimited responsibility for omissions is based on a misunderstanding of what an act is. Acts are 'homeostatic', i.e. they compensate for changes. They are focused on some end-result and away from others. For example, in football one keeps an eye on the ball and may not, because of this, see other things to be done. We cannot see the whole picture at once (Lucas, 1995, pp. 37-39).

I would like to support Lucas' argument with a reference to the theory of autopoiesis (from Gr. *auto* = self, *poiesis* = creation), developed by the cognitive scientists Humberto Maturana och Francisco Varela. According to this model of self-organization, each organism including each human being functions as an operationally closed system in the sense that external influences can disturb the organism but not specify the content of the organism's reaction. The reaction depends on the structure that the organism already has (Maturana, Varela, 1992, pp. 42-52, 94-103, 131-138).[2] This model can be used to support Lucas' notion of the homeostatic character of acts. Acts are ways of reacting to and coping with external disturbances. They are reactions to problem situations forced upon us.

This argument is of a conceptual nature. Other arguments are pragmatic. To place the responsibility for omissions on a complete equality with the responsibility for acts may produce an excessive sense of guilt. We can always say that we should have done more. This might be counterproductive–we feel guilty all the time but do effectively nothing about anything.

[2] For a presentation and analysis of this model, see Nordgren, 1994, pp. 80-82.

The asymmetry between the moral responsibility for acts and the responsibility for omissions has implications for the issue of responsibility in science. Scientists have a greater moral responsibility for what they actually do in their research than for their omissions.

Taking Responsibility

The different models of responsibility also have implications for the aspect of taking responsibility. Agential models neglect not only the fact that moral responsibility is a matter of ascription but also that it is a matter of assumption. The reason for this neglect is that the notion of assuming responsibility goes beyond the responsibility determined by causal agency. We may take responsibility even for acts and acts of omission which we have not caused. However, the proposed modified social model stresses not only the ascription of responsibility but also the assumption of responsibility. Since causal agency does not in itself determine moral responsibility, social groups or individuals might assume different responsibilities for themselves. In distinction to some social models, the model also stresses the possibility for individuals to assume a responsibility that deviates from responsibilities that prevail in their communities and social groups. At least to some extent, taking responsibility is a matter of individual choice.

Assuming responsibility is to take upon oneself the obligations and consequences of a situation. This can be done in different ways. In one sense, assuming responsibility is to accept responsibility when others hold you responsible. In the prototypical case, it is a matter of admitting that one has caused a certain act or consequence, that the act or consequence is morally relevant, that one is not ignorant of the act or consequence, and that one has not acted under coercion. As with ascribing responsibility, assuming responsibility may be backward-looking as well as forward-looking.

Another use of the phrase 'taking responsibility' concerns vicarious responsibility. It is possible to take on someone else's responsibility. Here it is not the case that one has caused an act or a consequence. Despite this, one accepts responsibility for it. An example is parents taking moral responsibility for something that their children, but not they themselves, are directly causally responsible for, such as breaking a window.

A third use of the expression 'taking responsibility' concerns assuming duties other than the ordinary ones. These may be more limited than the ordinary duties or more extensive. Lawyers, for instance, are not obliged to always speak the truth, as are citizens in general. The responsibility that lies with their social role is to defend their clients. And policemen are not obliged to avoid violence–in most circumstances–as are citizens in general. To their role as policemen is linked a right and even a duty to use force. Politicians, on the other hand, take a more extensive responsibility for the development of society than ordinary people do (May, 1992, pp. 165-166; Hart, 1968).

Having a Responsibility

'Responsibility' is often used as shorthand for 'obligation'. This means that it refers to a normative content of some kind, which can be understood in at least two different ways. Responsibility can mean 'obligation to act in accordance with a set of well-defined rules'. In this sense, responsibility is strictly rule-governed. But it can also mean 'obligation to make appropriate judgments in particular situations even when strict rules are not applicable, and to act in accordance with these judgments'. Here responsibility is not linked in a simple way to certain rules but includes an irreducible element of discernment. One might have responsibility for a certain area or domain, but it is not predetermined how one should decide or act in each particular situation. One must make a judgment from case to case (Jonsen, 1968, p. 69).

Sometimes obligation and responsibility are contrasted. One makes a distinction between 'ethics of obligation' and 'ethics of responsibility'. The former is matter of obeying rules set out in advance while the latter includes discernment. These two kinds of ethics, thus, correspond to the two meanings of 'responsibility' described above. I will come back to other normative interpretations of 'responsibility' in a later section.

Response and Responsibility

Certain modern philosophers and theologians focus on the listening-answering root meaning of the term responsibility. In other words, they stress the 'response' metaphor. Their models can be called encounter models of responsibility, focusing on the event of encounter between a human agent and some 'other'. In contrast to many agential and social models of responsibility, the encounter models are often presented as having certain normative implications.

We find a proponent of this view in the philosopher Emmanuel Lévinas. In his opinion, the starting point of morality is an encounter between two people where each one is obliged to respond to the presence of the other independently of whom or what the other is. That there is an obligation is clear from the mere presence of the other, while how one should respond is due to the particularities of the situation. In other words, ethics (obligation) precedes ontology (what kind of characteristics the other has). The 'face' of the other–a central concept of Lévinas–puts us under a moral obligation. The 'address' of the other demands a 'response'. This encounter is the ground of morality and the key aspect of responsibility (Lévinas, 1985).

Another exponent is the theologian Richard Niebuhr, to whom responsibility is a matter of being responsive to others. Responsible action is a response to action upon us. All organisms, including humans, interact with and respond to their environments. However, human responses are not mere reflexes or determined by instincts but actions based on interpretations of problem situations forced upon us. We answer the question "What is going on?" before we act. Moreover, humans can anticipate responses to their actions. This means, according to Niebuhr, that human agents are morally responsible for anticipated reactions to their actions. Ultimately, human life is lived in response to God (Niebuhr, 1963).

How should the encounter models be assessed in general? To be sure, response within the framework of an encounter is an important aspect of moral responsibility. However, it is important here not to conflate metaethics and normative ethics. Lévinas' and Niebuhr's models include both kinds. They stress the listening-answering aspect as a key to an understanding of the nature of responsibility, but they also give some general normative suggestions.

I find Lévinas' view, that ethics precedes ontology, unacceptable. It could be that we have obligations to the other independently of the person's particular characteristics. However, that the other is a person in the first place is also a matter of ontology. So, it seems clear that ontology precedes ethics logically as well as psychologically, not the other way around. Notwithstanding the difficulty of clearly defining a person, it is not self-evident that only persons have moral standing. Here, Niebuhr's model might give some guidance. By comparing human response to organisms' responses in general, he indicates a more promising approach. The different entities in the world actually exhibit a continuum of responses from mere reflex to full moral response. Conversely, it could be argued that human agents are responsible to all entities in the world with which they interact, not just to other humans. Moral responsibility, however, is present only in relation to certain entities (cf. Moran, 1996, pp. 59-65). Somewhere along the line mere responsibility, in the sense of responsiveness to their influence on us, turns into moral responsibility. But the border line is unclear and a matter of dispute. When does mere responsibility turn into moral responsibility?

This implies that the proposed 'modified' social model must be even more modified. Responsibility is not merely a matter within the human community. We have responsibility to nonhuman entities as well, although not always a moral responsibility. Responsibility is assumed and ascribed by human agents in response to other human and nonhuman entities. Social models of responsibility commonly include a recognition of encounters with other humans. But an acceptable social model should also include encounters with other biological entities.

Be Responsible! Act Responsibly!

The three kinds of models of responsibility discussed above–agential models, social models, and encounter models–can all be described as metaethical models. They describe the nature of moral responsibility–they give an answer to the question of what moral responsibility is all about. As such, they are possible to combine with many different normative ethical views.

However, it is also possible to give these different kinds of models normative interpretations. In these cases, responsibility is proposed as a first normative principle. The principle can be formulated as an imperative: 'Be responsible!' or 'Act responsibly!' What this imperative means more precisely depends on whether we talk about normative agential models of responsibility, normative social models, or normative encounter models.

As an example of a normative agential model, one may choose Immanuel Kant's ethics of autonomy. To be sure, Kant never used the term 'responsibility' in his writings but to the extent that autonomy, and thus responsibility for self, is central to his ethics, it can reasonably be interpreted as a normative agential model of responsibility. This is so because, as William Schweiker writes, "...the self-legislating freedom of the agent and principle of that legislation, the categorical imperative, is the first principle of ethics" (Schweiker, 1995, p. 81).

As an example of a proponent of a normative social 'model' I would suggest Marion Smiley. She not only describes the social nature of responsibility, she does so with certain normative goals in mind. By taking practical judgments about responsibility seriously, she hopes to use them "... to open up channels of communication with regard to the moral responsibility of particular individuals for the suffering of others" (Smiley, 1992, p. 14). It is important to note that Smiley does not attempt to develop a comprehensive normative model of responsibility. She merely hopes that a social model of responsibility will have some positive effects on the communication on moral matters. For this reason, I put 'model' within quotation marks.

However, among the normative models of responsibility, encounter models appear to be the most common. A great variety of thinkers have proposed a kind of normative ethics of responsibility starting from an experience of encounter. We have actually already met two of them, Lévinas and Niebuhr. They do not only describe the nature of responsibility, but also give certain general normative suggestions.

There are also, however, other kinds of normative ethics of responsibility than those that easily fit the scheme of normative agential, social and encounter models. One important example is Hans Jonas in his influential book *The Imperative of Responsibility*. This book is particularly interesting in our context, because Jonas means that modern science-based technology, and not least gene technology, brings the issue of responsibility to the center of the ethical debate.

In contrast to Lévinas, Jonas argues that ontology precedes ethics. Two ontological ideas are particularly important. First, Jonas views modern technology as an ontological event. It differs both quantitatively and qualitatively from earlier technologies. It has given immensely increased powers of humans to affect their own future as well as the future of the planet at large. One example is gene technology. Moreover, in Jonas' opinion modern technology threatens to endanger the existence of humankind. This brings us to his second ontological idea, namely an 'ontological proof' that humankind ought to exist. The traditional 'ontological proof' concerns the logical necessity of the existence of God. Jonas, on the other hand, tries to show that the continued existence of humankind is a moral necessity. Fundamentally, he tries to show that this continued existence has intrinsic value. Jonas views the responsibility of parents for their child as an 'archetype' of responsibility. The act of procreation affirms, at least implicitly, the value of continued human existence. The child is the face of the future for which we are responsible. According to Jonas, the experience of the child allows us to formulate an imperative of responsibility: Never endanger the existence of humankind! However, that is what modern technology threatens to do (Jonas, 1984).

I will not make any specific comments on these examples of normative models of responsibility. A general comment, however, is in place. Common to all, except Smiley's view, is a wish to ground moral thought in a fundamental single-valued model. I will argue below that such single-valued views are unsatisfying. Smiley's view, on the other hand, is acceptable but much too thin. It is actually not a normative model at all, only a hope. As a metaethical view, Smiley's social model is attractive. As a normative ethical view, however, it has very little to offer.

2. ENGINEERING OR 'TINKERING' IN MORAL REASONING?

So, what should an acceptable normative ethical view look like? What kind of a normative view should the proposed modified social view of moral responsibility be combined with?

There are two radically different approaches to practical moral reasoning and decision-making. The first approach can be called 'moral engineering'. Here morality is understood as a matter of principles. The principles are viewed as well-defined and literal, and they are applied to particular cases. I call this approach 'moral engineering' because moral reasoning according to this view is a matter of applying general ethical principles to specific moral problems in a way similar to that of an engineer applying general scientific principles to specific technical problems. The key element of morality is deduction from one or more ethical principles to particular cases. Therefore, the approach is deductivist in nature.

The second approach to practical moral reasoning and decision-making, on the other hand, is nondeductivist and can be called 'moral tinkering'. In distinction to an engineer, a tinkerer has only limited theoretical knowledge of science. Moreover, he does not plan his construction of new things in the careful way of an engineer. He takes material that happens to be in his disposition and uses his imagination to create new things. He is an inventor that uses old things in new ways without thorough planning and strict logical or mathematical reasoning. He may take a bicycle tube, a funnel and some other material, and make a new kind of machine using his imagination. Similarly, a 'moral tinkerer' does not start from ethical theories and abstract principles but from earlier moral experiences of more or less similar cases. In this kind of moral reasoning, moral imagination is the key element, not deduction from general moral principles. Moreover, moral concepts are not conceived of as well-defined and literal, but as fuzzy and metaphorical.

Moral engineering and moral tinkering represent two extremes. In modern bioethical reasoning we often find intermediate forms. They are responses to the complexity of many moral situations, and can be viewed as attempts of avoiding certain weaknesses of the pure 'engineering' approach and coming closer to the 'tinkering' approach. In my opinion, however, they are not entirely satisfying. Instead I shall propose a radical version of the 'tinkering' approach that I call 'imaginative casuistry'. My argument for this position takes as a starting point certain empirical results of cognitive semantics.

Cognitive Semantics

Cognitive semantics can be defined as the study of human concepts with empirical methods developed in cognitive science. It is common to distinguish between two generations within cognitive science. The first was dominated by a computer model of the mind. Human cognition, i.e., human thinking and reasoning, was understood as formal manipulation of symbols, and even if the mind was considered as being dependent on some kind of physical substrate, it was seen as being independent of any particular kind of physical substrate. One and the same mind could possibly exist in, for instance, a computer, a computer program written on paper, or a human brain.

Both these ideas are questioned by the second generation of cognitive science. First, it is maintained that cognition to a large extent has a metaphorical character. Often it is not a matter of literal and well-defined symbols. Second, the human mind is considered to be embodied in the sense that it gets its specific properties by its relation to the human body. The title of one of the books by Mark Johnson shows this very clearly: *The Body in the Mind* (Johnson, 1987). Human bodily experiences give an irreducible character to the human mind.

Cognitive semantics is a way of understanding human concepts from this bodily and, to a large extent, metaphorical perspective. Instead of trying to establish a priori the character of concepts, as is the case in traditional semantics, one takes as a starting point an empirical study of peoples' actual usage of concepts. Let us take a closer look at some of the general findings of cognitive semantics before we discuss its findings concerning moral concepts. Note that many of these ideas are shared with other types of semantics as well.

Concepts

The classical view in philosophy has been that conceptual contents can be explicated by specifying necessary and sufficient conditions. In order to specify the content in, for instance, the concept of a 'chair' one has to specify the set of properties which all chairs and only chairs have. In practice, however, it is often impossible to specify the precise conditions that determine exactly whether an object falls under a concept.

As an alternative to this view Wittgenstein talked about family resemblance (Wittgenstein, 1953, pp. 66-71; cf. Lakoff, 1987, pp. 16-17). According to this view an object A may resemble object B, and object B may resemble object C, without object C resembling object A. In this way, problems of demarcation become more intelligible. We understand why it so difficult to demarcate chairs, including kitchen chairs and rocking-chairs, from, for instance, bar stools, sofas, rocks one can sit on etc.

Another criticism of the classical view was suggested by the cognitive psychologist Eleanor Rosch in the 1970s. She presented convincing evidence for the idea that certain objects falling under a concept seem to be more typical than others. For example, a robin seems to be a more typical bird than a penguin or a pelican. The main point of her 'model of prototypes' is that certain objects seem prototypical, while others are nonprototypical. Concepts are radially structured with prototypical instances making up the center and nonprototypical instances radiating out at various distances (Rosch, Lloyd, 1978).

A further development was made by George Lakoff and Mark Johnson. In their book *Metaphors We Live By* (1980) and in later books and articles, they argue that not only do concepts exhibit prototype structure, many of them are metaphorical as well. Certainly, not all concepts are metaphorical, but metaphors are much more common than assumed in traditional semantics (Lakoff, Johnson, 1980, pp. 3-6).

Metaphors

Lakoff and Johnson maintain that the metaphorical understanding of our experience arises when we use concepts from one domain of experience (the source domain) to structure our experience in another domain (the target domain) (Lakoff, Johnson, 1980, p. 5; Johnson, 1996, pp. 51-52; Lakoff, Johnson 1999). A great deal of our thinking is metaphorical in this sense, but an important part is nonmetaphorical, and this is also an important idea in cognitive semantics. Nonmetaphorical concepts are those which refer to the body and to the direct interactions between the body and the world (Lakoff, Johnson, 1980, pp. 56-60, 115-125). It is very important to notice that the nonmetaphorical concepts are not more important, more valuable or more real than the metaphorical ones. They are only more basic in the sense that they are more directly linked to our bodily experiences.

However, a great deal of the conceptual contents are metaphorical. We imagine something within one domain of experience in terms of something from another domain of experience. For instance, we often use nonmetaphorical body-related concepts to understand abstract concepts. At other times we use metaphorical concepts to understand other metaphorical concepts.

An example of a metaphorical concept is 'argument' as used in academic discussions. We understand what an argument is with the help of concepts from other domains of experience. Among others the following metaphors are used: 'argument is war' ("Your claims are *indefensible*", "His criticisms were *right on target*", "He *attacked every weak point* in my argument", "I've never *won* an argument with him", "I *demolished* his argument"), 'argument is a building' ("With the *groundwork* you've got, you can *construct* a pretty strong argument", "If you don't *support* your argument with *solid* facts, the whole thing will *collapse*", "He is trying to *buttress* his argument with a lot of irrelevant facts, but it is still so *shaky* that it will easily *fall apart* under criticism"), 'argument is a journey' ("We will proceed in a *step-by-step* fashion", "We have *arrived* at a disturbing conclusion"), and 'argument is a container' ("Your argument doesn't have much *content*", "That argument has *holes* in it") (Lakoff, Johnson, 1980, pp. 4-6, 83-105).

A possible objection to this view on metaphors could be that it does not provide a clear criterion for which domain is the primary one when both the concept of the 'source domain' and the concept of the 'target domain' are metaphors. To this one might reply that in cases like these one must pay close attention to what words are actually used in giving content to a concept, and to the context of the utterance. The problem cannot be solved with logic. It can only be solved empirically. It is a matter of factual primacy, not of logical primacy.

In my judgment, the semantic position of Lakoff and Johnson seems well founded. For empirical reasons the prototype model is to be preferred to the view that concepts are defined by necessary and sufficient conditions. This is also true of the view that our concepts are to a large extent metaphorical.

Now, the question is what significance cognitive semantics has for the view on moral concepts. Johnson, especially, has studied this problem. His main contribution is the book *Moral Imagination* (1993).

Moral Concepts

Johnson argues that moral concepts are to a large extent metaphorical. This holds true of the most fundamental moral concepts as well as the concepts with which we characterize a situation. He also maintains that moral concepts are radially structured, i.e., exhibit prototypical and nonprototypical instances, and that moral reflection and debate often concerns whether or not to extend moral concepts to problematic, nonprototypical instances. Furthermore, Johnson asserts that moral reasoning concerning the extension from prototypical to nonprototypical cases involves imagination rather than deduction.

Let us take a closer look at these aspects.

Fundamental Moral Concepts

According to Johnson, the most fundamental moral concepts are metaphorical. Examples of such concepts are "... will, action, purpose, rights, duties, laws" (Johnson, 1993, p. 33). Johnson does not define more precisely what he means, but in my interpretation the most fundamental moral concepts are those that give an understanding of what morality is all about.

One important fundamental metaphor, or complex of metaphors, is the one that defines the correlative concepts of rights and duties. Johnson calls this metaphor the 'social accounting' metaphor. We owe things to others and others owe things to us. The 'social accounting' metaphor sets up a conceptual framework within which we understand our rights and duties. It does not, however, explain completely the procedure for assessing *what* we owe others and *what* others owe us. This is done via a version of the 'social accounting' metaphor, which may be called the 'moral interactions are commodity transactions' metaphor or the 'moral accounting' metaphor. Here the source domain is commercial exchange. Within this metaphor well-being is understood in terms of wealth, i.e., as access to valuable goods ("She has lived a *rich* life", "It is a *poor* man who has no love of himself", "He *profited* from his experience", "Nothing can compare to the *riches* of family, friends, and loved ones", "The cynics of the world lead *impoverished* lives", "Her improprieties *cost* her her job"). On the basis of the 'moral accounting' metaphor we reason about what is fair. We say such things as: "I couldn't possibly *repay* your kindness", "I *owe* you my life", "I'm holding you *accountable* for her suffering", and "In *return* for our kindness, she gave us nothing but grief". Duties are understood in terms of financial debt ("You are obliged to...", "It is your duty to...") and rights in terms of credit ("As a human being one has certain fundamental rights") (Johnson, 1993, pp. 35-50; Johnson, 1996, pp. 53-57).

Within this metaphor morality gets a predominantly reciprocal character. There is, however, some room for altruism as well. Sometimes people perform acts that go beyond what they owe others (Johnson, 1993, p. 49).

The 'morality is social accounting' metaphor is only one of many different metaphors used to define morality. Among other examples analyzed by Johnson one can mention 'morality is control' ("You have to *control* your urges"), 'morality is erectness' ("He has *sunk* very low"), 'morality is purity' ("She has been *purified* from her sins"), and 'morality is obedience' ("*Follow* the commandments of reason (conscience, God, the Law)!") (Johnson, 1993, pp. 50-52; Johnson, 1996, pp. 57-63). Other examples are 'morality is health', 'morality is politics' and 'morality is art'. Each one of these metaphors gives morality a distinct character. In the everyday moral talk we often use a mixture of several such fundamental metaphors.

Johnson maintains that all of our most fundamental moral concepts are metaphorical: "Our most fundamental moral concepts ... are defined metaphorically" (Johnson, 1993, p. 2). This is, of course, a strong statement, particularly since it is based on empirical evidence. But Johnson seems to believe that a strong inductive case can be made.

This view also applies to the concept of responsibility. As we have seen above, responsibility is also a metaphor. The term has been extended metaphorically in two directions corresponding to its two root meanings, 'answering' and 'accounting'. In the first case, we have an aural/oral metaphor. We listen and respond. In the second case, we have a financial metaphor. We owe things to others and others owe things to us. This latter metaphorical extension fits Johnson's analysis of morality in terms of the 'social accounting' metaphor particularly well.

It is obvious that these results are relevant to the issue of moral responsibility in science in general and human genetics research in particular. The 'response' metaphor stresses the importance of dialogue between scientists and society. The 'owing' metaphor stresses that scientists owe certain things to society because society has given certain credits to them in terms of education, funding, etc.

Framing of Situations

According to Johnson, concepts used to describe particular moral situations are also metaphorical. Traditional ethics has often neglected that the framing of a situation is metaphorical and that different ways of framing may have different implications for how we think and act (Johnson, 1993, pp. 53-54, 192).

One example of the metaphorical framing of situations given by Johnson concerns marriage. Marriage can be understood in terms of many different metaphors: 'Marriage is an investment', 'Marriage is an ongoing journey', 'Marriage is an organic unity', etc. These metaphors may have different implications for action. If marriage is viewed as an investment, it might easily be allowed to be broken up when business goes bad. If it is seen as an ongoing journey, it might also be allowed to be broken up easily. The car might stop functioning! If it is seen as an organic unity, on the other hand, it implies duties of monogamy, preserving unity and mutual growth (Johnson, 1993, pp. 53-61).

The importance of framing of situations is clearly seen in medical practice and scientific research. In a section below I shall discuss some examples of framing metaphors. It is obvious that framing metaphors such as 'medicine is war', 'medicine is business', 'medicine is machine repair', 'medicine is system maintenance', and 'medicine is personal care', have very different practical implications for medical practice and medical research.

Prototype Structure of Moral Concepts

Moral concepts, as concepts in general, are radially structured with prototypical instances making up the center and nonprototypical instances radiating out at various distances. The prototypes are clear, unproblematic, stable and widely shared. They are surrounded with unclear, problematic cases. These are the objects of moral reflection and are more or less disputed. The main problems in moral debate concern the degree to which concepts should be extended to such nonprototypical cases (Johnson, 1993, pp. 92, 189-192).

One example given by Johnson is the concept of a 'lie'. There are certain central instances of this concept that are generally and easily recognized as lies. This core, however, is surrounded by a large number of more problematic cases such as white lies, social lies, jokes, oversimplifications, mistakes, etc. There are different opinions about how far the concept of a 'lie' should be extended. Some people would argue that most of these cases are concepts in their own right. Others might argue that most of them can be included in the concept of a 'lie' (Johnson, 1993, pp. 91-98). Another example could be 'coercion'. There are clear, prototypical instances of coercion, as when someone threatens a person with a gun in order to make him do certain things. But what about economic pressure, social pressure, manipulation with rhetoric means: should these also be included in the concept of coercion? Some say 'yes', others 'no'. It is possible to extend the concept of coercion from prototypical to nonprototypical cases in many different ways.

Even concepts used in framing situations exhibit a radial structure. One important example is the concept of a 'person'. It is a widely accepted moral principle that we should respect persons. But what is meant by a 'person'? What instances fall under this concept? Johnson points out that white, adult, heterosexual and healthy males often have represented the prototypes of persons. They have been surrounded by nonprototypical instances such as females, children, fetuses, nonwhite people, homosexuals, and the mentally retarded. Moreover, it has also been heavily debated whether all members of the human species should be considered as persons, and if animals other than humans should (Johnson, 1993, pp. 97-98).

The prototype structure of moral concepts sheds new light on the debate between moral absolutists and moral relativists. The moral absolutist is right when asserting that there is a shared, stable, clear and unproblematic core. For these cases, it can be said that they fall under fixed moral principles. The absolutist is wrong, however, in generalizing this. There are unshared, problematic cases as well. Conversely, the moral relativist correctly stresses the existence of such unshared and problematic cases. But he or she is similarly wrong in generalizing this point of view, neglecting the existence of a shared, unproblematic core. The prototype model of moral concepts represents a middle course between moral absolutism and moral relativism (Johnson, 1993, p. 80).

The analysis of the concept of responsibility presented above indicates that this concept also exhibits prototype structure. There are prototypical instances of moral responsibility, but as Niebuhr's analysis shows somewhere along the line moral responsibility turns into mere responsibility. We have moral responsibility only in relation to some entities. It is a matter of dispute as to what extent moral responsibility should be extended to nonprototypical cases such as animals and ecosystems–there is a gray zone.

In a section below, I shall argue that the central principles in biomedical ethics and research ethics such as autonomy, justice, and integrity are also metaphors and exhibit prototype structure.

Moral Imagination

An important view in ethics is that morality has to do with principles. This view seems to presuppose that moral concepts are literal and well-defined, and that moral reasoning is a matter of deduction from moral principles to particular cases. However, Johnson argues that moral concepts are metaphorical and that moral reasoning has to do with imagination rather than deduction. He admits that principles play an important role in prototypical cases. In these cases it is a matter of deduction. But this is not true of nonprototypical cases. Here, it is a matter of using one's imaginative powers to make a metaphorical extension from prototypical to nonprototypical cases. This moral imagination consists of at least two different abilities. First, it is a capacity for empathetic imagination. Unless we can put ourselves in the place of others, we cannot be morally sensitive. Second, it is a capacity to envision various possibilities for action in a particular situation. For instance, our ability to criticize moral views depends on our capacity to imagine alternative views on specific moral cases (Johnson, 1993, pp. 198-203). To this I would like to add that the capacities of empathy and envisionment do not in themselves guarantee that we make the right decisions. There is always a risk that we might put ourselves in the wrong place, that we empathize with the wrong people. However, if we use our imaginative capacities persistently, for a long time and in dialogue with other people, this risk is minimized.

Johnson stresses that imagination functions in a narrative context. He uses the concept of a 'narrative' in a very broad sense. Prototypical narratives are stories that we tell others and sometimes write down. Johnson extends this metaphorically to include broad synthesizing structures within our very experience itself. He characterizes human beings as interpretive animals. Almost everything we do is done within a narrative understanding. This is also true of our moral reasoning. We may evaluate a person breaking a promise in one way in the narrative context of lying to his spouse about an extramarital affair, and in another way in the context of failure to keep his promise to an officer of a dictatorial police state. A vital aspect of moral imagination is exploration of different narrative contexts (Johnson, 1993, pp. 150-154).

As an example of the role of moral imagination in bioethics we can choose the current debate on selective abortion based on prenatal testing. Often, this problem has been stated in terms of personhood. With the terminology used in Johnson's model the problem understood in this way concerns whether the concept of 'person' should be extended to the unborn. Is a fetus a person or not? To what extent can a fetus be understood as a person? To solve this problem we must use our imagination. Deduction will be of little help. But perhaps the problem of selective abortion should not be stated in terms of personhood at all? Perhaps we should use our imagination to state it in another way? Possible alternatives could be that the problem should not be deliberated on in terms of the concept of personhood but in terms of the concept of parenthood and family values (cf. Murray, 1996, pp. 142-166), or in terms of women's liberty. These alternatives give the discussion a very different shape. The point is that solutions of problems such as selective abortion have to do with imagination rather than with deduction. Reasoning in bioethics is a reasoning in metaphors, and the main questions are not matters of deduction but of what metaphors we should use and how far we should extend them. Sometimes the solution of a problem in bioethics requires a change of metaphors.

Moral Theory

Johnson describes his theory as a 'moral theory'. What does this mean? What is it a theory about? Is it a semantic theory, a psychological theory, a metaethical theory, or a normative theory? Or something else? Johnson is not quite clear on this point. He says that one cannot separate moral theory from moral psychology (Johnson, 1993, p. 250), but what exactly is the relation? He also maintains that his findings are empirical (Johnson, 1993, p. 1). In that case, what is the relation between these findings and his moral theory? He is not quite clear on this point either. Below I will present a possible and, hopefully, plausible interpretation.

First, Johnson presents a semantic theory, i.e. a theory about concepts. He argues that the most fundamental moral concepts as well as the concepts framing moral situations are metaphorical.

Second, he develops a theory of moral psychology. This theory is intimately connected with his semantic theory. The metaphorical character of our basic moral concepts implies that moral reasoning is a reasoning in metaphors, and in such reasoning imagination is of fundamental importance.

Both his semantic theory and his theory of moral psychology are developed within the framework of cognitive science and particularly cognitive psychology. They are empirically founded theories about the ways human beings use moral concepts and how they actually reason morally. Their aim is to clarify what moral conceptualization is all about.

Third, Johnson goes beyond these empirically founded theories in asserting that morality in fact should be understood only in this empirical sense. In doing so he actually presents a metaethical theory, namely a fundamental theory about the nature of morality and moral reasoning. He tries to answer the question about what methods we should use to find out what to do or to be. I interpret this metaethical theory as naturalistic, because he answers the question with reference to empirical investigations about the ways human beings actually reason morally.[3] Within the framework of metaethical naturalism, his model can be more precisely classified as a kind of moral constructivism, i.e., as a model according to which what is right is an invention.[4] There is no such thing as the single 'right thing to do' that can be discovered. Certainly, from the perspective of the individual who is born into our species and our culture, moral conceptions are often perceived as pregiven. They are biologically or culturally inherited. From the perspective of the evolution of our species and the evolution of our culture, however, these conceptions are inventions. Furthermore, in other cases it seems that what is right is something that we construct in the present with our imagination.[5]

However, it might be possible to accept Johnson's empirical findings, his semantic theory, and his psychological theory, and at the same time reject his metaethical theory. One may believe in objective moral principles and at the same time accept that the limited human capacity of moral reasoning involves metaphors and imagination. But in my opinion such a metaethical theory does not take these findings seriously enough.

Fourth, Johnson's theory is not a normative ethical naturalism in the sense that it tries to answer particular normative ethical questions with reference to empirical knowledge about people's actual moral beliefs and behavior. It is a theory about methods, not about particular judgments. However, since Johnson defends a theory about methods, it is obvious that his theory gives some general guidance about how to live. According to his metaethical naturalism, an acceptable normative theory must be psychologically realistic in the sense that our normative moral views must be realizable by human creatures like ourselves. Our morality must be a *human* morality, a morality that it is possible for us to live up to and that is applicable to real situations in life. Moreover, knowing oneself and other human beings can help us to understand problematic situations and find constructive solutions (Johnson 1993, pp. 184-189; Johnson, 1996, pp. 48-49). Johnson's metaethical theory has at least some implications for normative ethics, but it does not tell us how we should act in particular situations.

I accept Johnson's overall position. His arguments seem convincing. Now, I will develop the normative implications with special reference to bioethics and research ethics.

---

[3] Cf. May et al. (eds.), 1996, pp. 1-4. See also Johnson, 1996, pp. 63-67.
[4] Note that there are other kinds of metaethical naturalism that are not constructivist, but objectivist.
[5] For a more extensive discussion of this kind of view, see Nordgren, 1994, pp. 153-156.

3. IMAGINATIVE CASUISTRY VS. OTHER NORMATIVE POSITIONS

Johnson's moral theory seems to demand a radical change in traditional normative theory. In this new conception principles and deduction do not play the central role they do in the 'moral engineering' approach. Instead, the focus is on metaphorical extensions from prototypical cases to nonprototypical cases with the help of imagination. This normative view clearly belongs to the 'moral tinkering' approach and can be more precisely classified as a kind of casuistry. The main argument for this view is that it seems to fit in very well with the empirical findings of cognitive semantics. In the light of these findings, it is a psychologically realistic kind of normative ethics.

However, before I develop this kind of normative position, I shall briefly discuss some other normative ethical views: first, Kantianism and utilitarianism; secondly, 'principlism' and 'reflective equilibirum'; and, thirdly, three versions of casuistry that have been particularly important in modern bioethics, namely the approach proposed by Albert Jonsen and Stephen Toulmin, the view advocated by Baruch Brody, and Carson Strong's version. After my presentation of these ethical views as well as my own proposal, I shall finally compare this proposal with another position, namely virtue ethics.

Kantianism and Utilitarianism

Kantianism and utilitarianism are two very influential normative ethical views. At least some versions of them can be viewed as examples of the 'moral engineering' approach, namely those according to which the theories are intended to give moral justification as part of a decision procedure. Moral justification in this sense is to provide good reasons for a particular ethical judgment or an ethical rule. However, it is also possible to view Kantianism and utilitarianism as theories of justification in a sense that is independent of the decision procedure. In this case, Kantianism and utilitarianism are thought to provide criteria of what is right, not a method of how to proceed in practical moral decision-making. If one holds such a view of moral justification, it might be possible to combine Kantianism and utilitarianism with the 'moral tinkering' approach to decision-making. In my opinion, however, this view is not satisfying. It does not take the empirical findings of cognitive semantics seriously enough. A view of justification not in line with these findings is not psychologically realistic. Below, however, I shall focus exclusively on Kantianism and utilitarianism as theories that are supposed to provide moral justification as part of a decision procedure, not as a separate issue.

The method of moral justification used in these decision-oriented versions of Kantianism and utilitarianism is deductivism. Morality is understood as a matter of principles. The principles are seen as well-defined and literal, and are applied by way of deduction to particular cases (cf. Jecker, 1997, 118-119).

The Kantian principle can be articulated in terms of respect for persons or respect for human dignity. This idea is expressed in the 'categorical imperative' that we should treat other persons always as ends in themselves and never as means only. Another version of the categorical imperative focuses on universalizability. We should act only according to that maxim by which we at the same time will that it can become a universal law.

The utilitarian principle, on the other hand, is to maximize expected utility.[6] Utility can be interpreted in terms of, for instance, pleasure, happiness, or preferences. In act utilitarianism, one seeks to determine in each case which act would maximize expected utility. In rule utilitarianism, one focuses on the rule or set of rules the implementation of which would maximize the expected utility.

The weaknesses of Kantianism and utilitarianism are well-known. A common objection to Kantianism is that it is unable to handle situations in which one has obligations to two or more persons and cannot treat all of them as ends in themselves. An example from a clinical context of such conflicting duties is when parents refuse aggressive life-preserving treatment for children with serious birth injuries (Strong, 1997, p. 68). Respect for the parents would imply that their autonomy should be respected. Respect for the children would imply that they should be given the treatment.

A common criticism of utilitarianism is that it underestimates the difficulties in calculating and comparing expected utilities. An example from medical practice is when women refuse treatment for the sake of their fetuses (Strong, 1997, p. 67). A rule utilitarian interpreting utility in terms of happiness faces an extremely complex situation. There are not enough medical, psychological and sociological data available to reach a well-founded decision. We do not know for sure the degree to which treatment would save lives and prevent neonatal handicaps. We do not know the psychological effects on the women of forcing them to undergo treatment, or the psychological effects on the health professionals. Moreover, we do not have sociological data enough to predict the long-term societal outcomes of forcing the women. Neither scientific data nor common sense give guidance enough. A consequence of the uncertainty and complexity is that it becomes extremely difficult to choose between the rule of always giving priority to the well-being of the fetuses and the rule of always giving priority to the autonomy of the women.

For similar reasons, it becomes extremely difficult for the act utilitarian to make such choices in each new situation. The problem with respect to each alternative act

[6] The expression 'expected utility' indicates that utilitarianism here is interpreted as providing a decision-method. As mentioned above, utilitarianism can also be understood as providing only a criterion of what is right. In this case, not the expected, but the actual consequences are referred to. Because we can never have knowledge of all consequences as compared with all consequences of all alternative actions, we can never, according to this view, know whether an action is right. Defenders of this view, however, do not see this as an argument against the plausibility of the view. Utilitarianism as a theory of justification is one thing, utilitarianism as a decision-method quite another.

of predicting and comparing the happiness of the parties directly involved as well as indirectly affected should not be underestimated.

To these common objections I would like to add other criticisms based on the empirical findings of cognitive semantics. It is obvious that Kantianism and utilitarianism, with their emphasis on well-defined ethical principles and deduction, do not fit the empirical findings of Johnson. The moral engineering approach of these ethical theories is useful in prototypical cases, where morality is a matter of deduction from principles. But such an approach is of limited use in the efforts of establishing prototypical cases, as well as in nonprototypical cases such as conflicts of duties or cases with vast complexity. Morality is a matter of metaphors and imagination to an extent that Kantians and utilitarians do not commonly recognize and probably cannot adapt to.

However, the question naturally arises as to why so many philosophers–Kantians, utilitarians, and others–have failed to notice the metaphorical character of morality. Can they really all be mistaken? To this, one can only answer that it is clear at closer examination that even the language used by these philosophers is in fact metaphorical, although they do not always seem aware of it themselves. Lakoff and Johnson examine at length the writings of Kant and find that this is actually the case (Lakoff, Johnson, 1999, pp. 415-439). Johnson also discusses early utilitarianism and comes to a similar conclusion (Johnson, 1993, pp. 120-123). This is not, however, the proper place to discuss their analyses.

In order to illustrate the difficulties of combining Kantianism, as it is often understood, with the empirical findings of cognitive semantics, let me quote Lakoff and Johnson (Lakoff, Johnson, 1999, pp. 438-439):

> Every aspect of second-generation cognitive science is at odds with the account of reason that Kant requires. What this means is that empirical results about the nature of mind can contradict philosophical theories of morality. Cognitive science presents us with an "is" that can contradict an "ought." When this happens, we maintain, we must opt for the most cognitively realistic position that is supported by the widest range of converging empirical evidence about the nature of mind.

What Lakoff and Johnson criticize here is Kant's idea of a 'pure practical reason' that is nonmetaphoric and disembodied. They find this view psychologically unrealistic. They stress, however, that this does not mean that Kant has nothing to teach us about morality. Far from it. Their point is only that we cannot found morality on his conception of reason. There is no such reason according to the findings of cognitive science (Lakoff, Johnson, 1999, p. 439).

Let me end this section by emphasizing that Kantianism and utilitarianism, to be sure, have several valid points. The Kantian principle of respect for human dignity is an important ethical principle, and consequences do matter ethically, as utilitarianism states. What I have been criticizing here is Kantianism and utilitarianism to the extent that they are examples of the 'moral engineering' approach to decision-making.

Principlism and Reflective Equilibrium

Another important view in modern bioethics is the four-principles approach or 'principlism' proposed by Tom Beauchamp and James Childress (Beauchamp, Childress, 1994 (1979); cf. Gillon, 1986). Principlism focuses on 'midlevel' principles, i.e. principles at a level between ethical theory and ethical rules, with particular judgments even further down in the hierarchy of ethical norms. The four well-known principles are autonomy, beneficence, nonmaleficence, and justice.

Principlism might seem to presuppose that basic moral concepts are literal and well-defined, and that moral reasoning is fundamentally a matter of deduction from moral principles to particular cases. However, this would be an inadequate description, at least of its present-day version. Childress is one of those who have contributed to the increased awareness of metaphors in bioethics in recent years. This might indicate that principlism is compatible with a metaphorical understanding of morality. Moreover, the importance of achieving a 'reflective equilibrium' between ethical principles and particular moral judgments has recently been stressed by Beauchamp and Childress. In addition, the principles are understood as prima facie principles, not as absolute, and a balancing between conflicting principles is often considered necessary. All this indicate that their approach is nondeductive. Let us take a closer look at the aspect of metaphor as well as that of reflective equilibrium.

With regard to metaphors, one tempting option would be to interpret Childress as focusing on the rhetorical use of metaphors. By 'rhetorical' use, I mean use of metaphors as a tactical device to bring home certain points in an argument. These points might be promotional, critical or pedagogical. However, Childress explicitly maintains that metaphors are not merely decorative or affective. They are central to both discourse and practice. Moreover, he argues that they are not necessarily incompatible with a principle-oriented approach (Childress, 1995, p. 1772).

However, I only partly agree with Childress. I think that it is possible to combine Johnson's model with principlism, if the principles are understood in terms of metaphors and metaphorical extensions. However, as far as I can understand, this would make principlism rather different from how it is usually described. As far as I can see, Childress has not taken the metaphorical character of ethical reasoning seriously enough. He has made valuable contributions to the understanding of the role of metaphors in bioethics, but he has not revised his principlism radically enough to fit in very well with the findings of cognitive semantics.

What about Beauchamp's and Childress' view on the relation between principles and particular judgments? If we compare the third and fourth editions of their book *Principles of Biomedical Ethics*, we find an interesting development in their thought. In the third edition, their approach is explicitly deductive. Particular judgments and actions are justified by rules, which are justified by principles, which are justified by ethical theories (Beauchamp, Childress, 1989, p. 6). However, in the fourth edition, they explicitly defend a view that is "neither top-down nor bottom-up; it moves in both directions" (Beauchamp, Childress, 1994, p. 20). This means that they reject both deductivism (top-down) and inductivism, i.e. the view that judgments on particular cases justify ethical principles (bottom-up). They call their view 'coherentism' and stress that the goal of this form of justification is a 'reflective equilibrium' between ethical principles and particular judgments. Reflective equilibrium allows for the possibility of modifying general as well as particular judgments. It should be noted, of course, that the idea of 'reflective equilibrium' has a history quite independent of principlism. This kind of justification was suggested by John Rawls in his well-known book *A Theory of Justice* (Rawls, 1971). In fact, reflective equilibrium has developed into an alternative position of its own in bioethics with proponents such as Norman Daniels, John Arras, and others. Norman Daniels suggests a 'wide reflective equilibrium' also including background theories regarding, for example, human nature (Daniels, 1979; Arras, 1991; Willigenburg, Heeger, 1991).

Beauchamp's and Childress' move away from deductivism is attractive. However, it is still not clear how far they are prepared to go in modifying their general principles. How much of the bottom-up approach is actually allowed for? If they are prepared to adopt reflective equilibrium in its full sense, it is difficult to see how their view can still be described as principle-based. Moreover, even if they would be prepared to adopt reflective equilibrium in its full sense, it is not clear that this position as such is satisfying. To be sure, the idea of reflective equilibrium in ethics has certain weaknesses. From the point of view of cognitive semantics, it can be criticized for not completely acknowledging the inductive aspect, i.e. that the specific interpretations of the ethical principles are justified by prototype cases. Moreover, it does not fully recognize the metaphorical character of the principles and metaphorical extension as a particular form of inductivism.

In conclusion, present-day principlism is much better than the older version, but it is nevertheless not entirely satisfying; neither is reflective equilibrium, as a position of its own. Let us therefore take a look at three more promising alternatives, namely the casuistry of Jonsen and Toulmin, the casuistry of Brody, and the casuistry of Strong. Perhaps these inductive approaches better fit the findings of cognitive semantics.

Consensus-Based Casuistry

Casuistry, i.e. a case-by-case approach to ethics, had its hey-day in the 16th and 17th centuries. However, it has had a rather low status in traditional academic philosophy. On the one hand, it has been criticized for being too rule-oriented in the sense that it includes a vast number of specific maxims for particular situations. On the other hand, it has been criticized for being too lax, for allowing almost anything depending on the circumstances. To the extent that this criticism is appropriate, this use of the casuistic method represents an abuse. This is at least the conlusion of

Albert Jonsen and Stephen Toulmin in their important book *The Abuse of Casuistry* (1988). In their opinion, however, an adequate use of the cauistic approach has much to offer.

Jonsen and Toulmin discern the following elements in this classical casuistic method.

(1) Paradigm and analogy. The moral cases are ordered taxonomically from the clearest and simplest, the paradigm cases, to the more complex and difficult. The transition is made by way of analogy.

(2) Maxims. In the assessment of individual cases, moral maxims of different kinds are referred to. These function as inference principles of the moral arguments. The maxims are more or less general, but never universal.

(3) Circumstances. By adding complicating circumstances to the paradigm cases, cases of increasing difficulty are construed. The traditional list of circumstances includes the following aspects: who, what, where, when, why, how, and by what means. It is the circumstances that 'make' the case.

(4) Probability. The cases are qualified in terms of the probability of the conclusions, from very certain to very uncertain conclusions.

(5) Cumulative arguments. A conclusion seems more or less probable, not on the basis of strict logic, but from the accumulation of reasons of varying kinds.

(6) Decision. The analysis of a case ends up in practical advice concerning the particular problem (Jonsen, Toulmin, 1988, pp. 250-257).

In their modern application of this classical casuistry, Jonsen and Toulmin emphasize that practical moral arguments primarily aim at establishing paradigm cases. These are normative in the absence of complications. The problematic cases are of three kinds. The first consists of marginal cases where a paradigm is not applicable in an unambiguous way. The second type occurs when two or more paradigm cases, in a contradictory way, are applicable to the same case. The third type consists of problems that arise when historical changes–psychological, social, technological–undermine the factual presumptions of our former moral concepts and views and force us to rethink our position in a fundamental way. In all problematic cases judgment and wisdom are needed to carry out an adequate balancing of general, but not universal, principles (Jonsen, Toulmin, 1988, 322-332).

A strong aspect of Jonsen's and Toulmin's model is their idea of paradigm cases. These seem to correspond almost exactly to Johnson's 'prototypical cases'. Furthermore, the conception of marginal cases seems to fit in very well with the idea of nonprototypical cases. Certainly, even the cases of conflict and the historically new cases may be interpreted as such nonprototypical cases.

Another strong aspect is their emphasis on analogy–one inductive method among others–instead of deduction. This is in line with the talk of cognitive semantics about metaphorical extensions. However, there are some differences between analogies and metaphors that are important to point out. This is particularly so, since this point constitutes an important difference between my version of casuistry (see below) and the casuistry of Jonsen and Toulmin. One difference between analogies and metaphors is that analogies draw on established similarities, while in doing metaphorical extensions we create similarities. Another difference is that in making analogies we are making a kind of inference, while with regard to metaphors we may use our imagination much more freely. One further difference is that in making analogies we can use literal concepts, while metaphorical extensions by their very nature are nonliteral. In sum, Jonsen and Toulmin seem to neglect the metaphorical character of the moral concepts themselves. This neglect of the metaphorical character of moral reasoning is a disadvantage of their view.

Another disadvantage is their overestimation of the consensus in modern Western societies. Some critics have argued that their version of casuistry requires social consensus about what constitutes paradigm cases. Such a consensus may have existed in the 16th or 17th centuries, but, they complain, is lacking in our pluralistic society, and since it is lacking, the method will not be of any help in practical life (Arras, 1991, pp. 42-44; Wildes, 1993, pp. 39-40). I am not prepared to go as far in my criticism, but nevertheless I think that the disagreement is greater than Jonsen and Toulmin suggest. There is some consensus, but probably not as large as they assume.

Finally, there is the weakness of stressing too hard the particularity of individual cases. This kind of casuistry does not seem very well suited for issues of broad social concern. Its usefulness in clinical ethics is apparent (Jonsen et al., 1998 (1982)), but is it truly a good starting point for policy-making? I doubt. A case-by-case approach is less appropriate in matters of policy than in clinical judgments. Broad social concern might demand fixed policies.

Intuitive Casuistry

Another kind of casuistry has been proposed by Baruch Brody (Brody, 1988; 1998). Brody argues that different views such as utilitarianism in its various forms, deontological models, natural rights models, virtue models and social contractarian models each has a moral appeal whose legitimacy is unquestionable. But at the same time each has failed because it has focused on only one of many legitimate appeals. Consequently, the legitimacy of each of these moral appeals is limited. Therefore, ethical problems have the character of balancing moral appeals, which in themselves are completely acceptable but may be in conflict with each other in particular cases. The results of such balancing are confirmed with reference to intuition.

According to Brody, the process of moral decision-making consists of three stages.

(1) The first stage is to make intuitive judgments of particular problems or situations.

(2) In the next stage we formulate a 'theory' about these initial judgments. These 'theories' can be simple generalizations covering only a very specific problem or complex theories covering a large range of issues.
(3) The final stage is a reconciliation between theory and initial judgment. To the extent that the model produces results that seem intuitively acceptable, the theory is confirmed.

At each stage the results are tentative and open to revision. Some of the initial intuitions might need to be rejected, and the 'theories' might need to be revised because of new intuitive judgments in new situations (Brody, 1988, pp. 9-14).

I would like to classify Brody's view as an 'intuitive casuistry'. It is a 'casuistry' because it is case-oriented and because it tries to weigh against each other different often conflicting but nevertheless in themselves acceptable principles. It is an 'intuitive' casuistry because it starts from intuitive judgments and because the results of the weighing are confirmed with reference to intuition. Brody himself prefers to use the label 'pluralistic' casuistry (Broady, 1998, p. 205). This is due to its emphasis on a legitimate plurality of different ethical principles that are not derivable from a single fundamental principle or value. For my purposes, however, the central role of intuition is more important to highlight in a classification than the acceptance of multiple principles.

A strength of Brody's model is that he stresses the legitimacy of many different moral approaches as well as the limitations of each one. There are many legitimate principles and values but in practice there are always problems in determining their scope of applicability as well as balancing them when they conflict. Another strength is that he refuses the idea that moral decision-making in general is a matter of deduction. By talking about intuition he recognizes that decisions in difficult cases are based on moral sensitivity.

However, a problem with Brody's casuistry is that his idea of intuition is intimately related to the view that what is right is something to be discovered, something objectively given. As mentioned earlier, Mark Johnson's theory seems to imply that what is right is something that is invented, and, in my opinion, this view is to be preferred. Brody refers to convergence of ethical views on particular issues as an argument for an objectively given morality (Brody, 1998, pp. 208-212). To this I have three objections. First, convergence should not be overestimated. To be sure, there is general consensus in many cases, but pluralism and lack of consensus are also apparent. Second, convergence can as easily be accounted for by a constructivist approach. People tend to have the same opinion in some cases (prototypical cases), and diverge in others (nonprototypical cases). Third, rather than being a sign of objectivity, convergence can reasonably be viewed as an indication of moral soundness. A view is morally sound if it is intersubjectively acceptable.

Another weakness of Brody's position is that he does not explicitly recognize the metaphorical character of the moral concepts, but seems to believe that they are quite literal and well-defined. Consequently, Brody's view does not fit in well with the empirical findings of cognitive semantics.

Policy-Permitting Casuistry

Finally, I shall discuss the casuistry of Carson Strong (1997). In a way quite similar to Brody, Strong starts from a recognition of the legitimate plurality of values and principles. His argument for his specific version of casuistry focuses on the problem of assigning priorities within this legitimate plurality of values and principles. He discusses four possible approaches. The first could be called general hierarchical ranking. Here the principles and values have a fixed order that holds for all cases without exception. One example is Robert Veatch's 'lexical ordering' where deontological principles as a group rank higher than consequentialist principles and values. The second approach is hierarchical ranking regarding particular issues. According to this view, the ordering holds for all cases in which a given issue arises. Examples could be: "Abortion is always wrong", "Surrogate motherhood should never be permitted" and "Germline gene therapy should never be allowed." The third approach discussed by Strong is what he calls 'casuistry' and exemplifies with Jonsen's and Toulmin's version. Casuistry is characterized by a case-by-case ranking. Here, there is always a possibility for exceptions due to contextual factors. Strong finds this flexible approach attractive. It avoids the oversimplification of the first two approaches and better reflects how decision-making in bioethics usually does and should take place. A weakness, according to Strong, is, however, that it nevertheless falls short of the flexibility desirable in dealing with the complexities of bioethics (Strong, 1997, pp. 71-75). He argues that (Strong, 1997, p. 75):

> [a]lthough generally cases should be decided individually, it is possible that for some issues there might be compelling reasons to prioritize similarly in all cases.

He continues (Strong, 1997, p. 75):

> ... for some issues there might be broad social considerations that provide reasons for adopting a uniform policy across all cases.

An example is the policy on reproductive issues that Strong himself defends, namely that physicians should refuse all requests for prenatal genetic testing for nondisease characteristics (Strong, 1997, pp. 75, 150-151).

Strong calls his version 'modified casuistry' because it allows for fixed policies for some issues and does not demand a 'strict' case-by-case approach (Strong, 1997, p. 75). For my purposes, a better label is 'policy-permitting' casuistry. This makes the difference between this form of casuistry compared to the other ones discussed in this chapter more clear than Strong's own designation.

I find Strong's 'policy-permitting casuistry' very attractive. Communal decision-making, whether in political or scientific communties, requires that, regarding specific issues, we sometimes prioritize between principles and values similarly in all cases. Policies are often socially desirable and if casuistry is to be useful in such a context, a strict case-by-case approach must be avoided. However, I do not agree with Strong that the casuistry of Toulmin and Jonsen is incompatible with such an view, so that it must be modified. On the contrary, their casuistry does permit that cases that are similar be treated similarly, and the extent to which such similarities exist can only be discerned in practice. This means that Strong's version of casuistry does not necessarily need to be called 'modified'. However, since he so explicitly stresses the importance of a casuistry that permits policies, there is good reason to call it 'policy-permitting casuistry'.

A weakness of Strong's casuistry, however, is that it does not explicitly recognize the central role of metaphors and moral imagination. This role is not least important in policy-making. Accordingly, Strong's view does not fit the empirical findings of cognitive semantics very well.

Imaginative Casuistry

As an alternative to consensus-based casuistry, intuitive casuistry, as well as policy-permitting casuistry, I would like to suggest an 'imaginative casuistry'. This normative ethical position preserves the best from all these versions at the same time as it takes the empirical findings of cognitive semantics seriously. The main point of this version of casuistry and what primarily distinguishes it from other versions is that it explicitly recognizes that moral reasoning is a matter of imagination. One makes metaphorical extensions from prototypical to nonprototypical cases.

The strength of Jonsen's and Toulmin's casuistry is its emphasis on paradigmatic (prototypical) and nonparadigmatic cases, and on analogies instead of deductions. The strength of Brody's version is that he recognizes the legitimate plurality of ethical principles and the limits of moral deductions. The scope of application and the priority between different principles and values have to be determined from case to case. The strength of Strong's casuistry is that, with regard to specific issues, it explicitly permits policies across all cases.

The imaginative casuistry also emphasizes the occurrence of prototypical and nonprototypical cases, and recognizes the limits of moral deductions. In both these respects it also takes the empirical findings of cognitive semantics seriously. This is true of its emphasis on the role of imagination and metaphors in morality as well. It also accepts that uniform policies are sometimes necessary. Moreover, putting emphasis on prototypical cases and imagination does not imply that ethical principles become redundant. Principles still have a very important function. They remind us about important considerations that should be included in our moral deliberation and are therefore often good starting points in ethical reflection. They are rules of thumb summarizing the collective wisdom regarding prototypical cases, not technical rules which tell us exactly what to do in specific situations (cf.

Johnson, 1993, p. 256). By 'collective' wisdom I mean that of the Western society, and the principles I refer to are those such as autonomy, utility, justice and integrity. I accept this plurality of principles as legitimate because they are all based on the collective moral experiences of our society. At an abstract level they seem to be generally accepted. For example, they can all be found, more or less explicitly, in ethical guidelines for biomedical research and clinical practice.[7] Taken at face value they seem uncontroversial in our society.

In imaginative casuistry, however, principles such as these are explicitly recognized as metaphors. Below, I will argue that they are metaphorical, and that these metaphors have been extended in many different ways. Moreover, the interpretive contents of the principles are determined by what are considered to be prototypical cases. In this sense the approach is fundamentally inductive. Of course, once the content is determined by the prototypical cases, the prototypical cases can be deduced from the principles, as is often done. But basically, it is the other way around. The important thing is from where the content fundamentally comes, on what basis it is ultimately justified, not the frequency in practical discourse of inductive and deductive approaches (cf. Winkler, 1993, pp. 361-362).

But not only can the principles be interpreted in different ways, they can also be assigned different scopes of application, and be balanced differently when they conflict. This makes it obvious that imaginative casuistry does not provide clear-cut normative solutions to ethical problems like single-valued theories, such as Kantianism or utilitarianism, or multi-valued theories with a hierarchical ranking of values or principles. It only presents a general decision-method that can be used in different ways, and a theoretical framework for this method.

How should the method of imaginative casuistry be described? How does one work oneself through an ethical problem on the basis of this ethical position? I would like to suggest the following procedural steps.

(1) Describe the problematic situation (case), identify the metaphors framing it and its nonprototypical characteristics.

(2) Identify relevant ethical principles such as autonomy, utility, justice, and integrity. Identify the prototype cases that support your interpretations of these principles and make explicit their metaphorical character.

(3) In deliberating on the case, use a wide range of imaginative methods, for example:

- imagine different framings of the situation (case),
- use your empathy to explore the experiences and feelings of other people involved in or affected by the problem situation,
- envision and criticize different alternatives of action,

[7] See, for example, the *Helsinki Declaration* (1964, 1975, 2000). It is published, for instance, in Annas, Grodin, 1992, p. 331-336. See also URL: www.wma.net.

- imagine different interpretations, applications, and rankings of ethical principles implicitly or explicitly referred to in different arguments in favor of as well as against the different alternatives of action,
- examine the implications of the metaphors behind the principles, and
- in matters of policy-making, take into account aspects of broad social concern.

(4) If possible, participate in a dialogue with other people. See to it that many different views are represented.

(5) Take a stand with regard to the interpretation, application, and ranking of relevant ethical principles. Make metaphorical extensions from the prototype cases or change the metaphors.

(6) Make a decision, whether in individual cases or in policy-making, in awareness that different persons may end up in different conclusions.

A possible objection to this kind of casuistry focuses on the risk of arbitrariness. This is, of course, a common objection to all kinds of casuistry, but it might hit this version to a larger extent than others. Ethics becomes a matter of imagination and it could be thought that one might imagine almost anything. Is not morality just the opposite of this?

However, this objection is only partially justified and I will reply to it in three steps. First, this is a criticism that could hit all kinds of ethical models. One always starts from unproven assumptions. Are there really any absolute foundations? Moreover, imaginative casuistry does not claim to reach certainty in its conclusions.

Second, there is nothing inherently wrong with imagination and freedom. It is something that characterizes human existence. It is something we have to live with for better or worse.

Third, there are certain limits to arbitrariness. Johnson stresses that objectivity in the sense of 'transperspectivity' is possible. Transperspectivity is a psychologically realistic *human* objectivity, in distinction to the impossible God's-eye-view objectivity. With regard to morality, transperspectivity is the ability of a biologically and culturally situated self to reflect critically on its own moral constructions and imagine other possible constructions. It might seem strange to combine objectivity and imagination in this way, but it is, in fact, imagination that makes human objectivity possible. Moreover, transperspectivity also has a social aspect. It is worked out by a community over an extended period of time (Johnson, 1993, pp. 240-243).

Another possible objection to imaginative casuistry is that it does not truly solve any normative ethical problems. It might have some clarifying points, but it does not give any clear answers to specific normative problems. However, according to imaginative casuistry, morality is not a matter of applying a fundamental single-valued theory to a particular problem and ending up with a clear-cut solution. There are no short-cuts in ethics. We have to interpret and balance the relevant principles in each case or type of cases, and the result depends, in part, on which cases we choose as prototype cases, i.e. on our previous experience of problem-solving, and in part on how we carry out the metaphorical extensions to nonprototypical cases. This means that different persons may come to different normative conclusions. All we can hope for is that we reach some consensus. Moreover, since transperspectivity is central to imaginative casuistry, narrow-minded views are excluded. Imaginative casuistry implies that we should empathize with other people and take into consideration many different perspectives, alternative actions, principles and values. This makes certain views improbable as results of ethical deliberation.

A third possible objection is that defenders of imaginative casuistry may end up in the same normative views on particular bioethical issues as those who defend other, perhaps more well-known, ethical theories. What is presented as a new ethical framework is perhaps not so new after all, at least not in practice. My reply to this objection is simply that the fact that people starting from divergent ethical theories end up with the same conclusions is a good thing. This does not imply, however, that imaginative casuistry is not a new and, perhaps, important position in ethical theory.

With these points in mind, we can now summarize the normative status of imaginative casuistry. It is obvious that imaginative casuistry is less directive than single-valued theories such as Kantianism and utilitarianism. It is also less directive than multi-valued theories with a hierarchical ranking of values and principles. To be sure, there are certain normative restrictions in terms of 'transperspectivity' and empathy. But it is clear that different persons using this method may end up in divergent conclusions. Imaginative casuistry is an instrument that can be used in different ways. Each individual has to make his or her interpretation, application and ranking of the ethical principles. As a result of dialogue, however, some consensus may emerge, at least on some issues.

This limited directiveness may be viewed as a weakness of imaginative casuistry as a normative position, but it actually makes it very appropriate in bioethics and research ethics. Together with a focus on the concept of responsibility, it may provide a common platform for dialogue within society and the scientific community in a way that other more strictly directive positions can never accomplish (see p. xvii).

Imaginative Casuistry vs. Virtue Ethics

In recent years, there has been a renaissance of the 'virtue ethics'. Starting in pure academic philosophy with Alasdair MacIntyre (1985 (1981)), Martha Nussbaum (1986), and others, it has spread to applied ethics, not least bioethics. Several authors argue that what is really needed by physicians in clinical practice is the ability of wise decision-making, not theoretical knowledge about abstract ethical principles (Toon, 1993). Above I have argued for a normative position that I call imaginative casuistry. This view also criticizes the focus on principles. The question arises: What is the relation between imaginative casuistry and virtue ethics? In order to answer this question we have to clarify the concept of virtue ethics.

Most important, virtue ethics is character-oriented, not action-oriented as are the other views presented so far, including imaginative casuistry. The key idea is that a good person performs right actions, not that a person who performs right actions is good. However, what is a good person? It is one who has certain desirable traits of character. To be sure, there are many different views on what these are. Common examples in bioethics and research ethics are integrity, honesty, and empathy.

In the traditional Aristotelian virtue ethics, the traits of character are thought of in terms of human flourishing and 'excellence'. Important examples are temperance and courage. Even more important, however, is practical wisdom. Aristotle clarifies what wisdom, *phronesis*, is by distinguishing it from *episteme*. *Episteme* is theoretical knowledge, for instance, geometry. Here statements and arguments hold true universally and with necessity. *Phronesis*, on the other hand, is characterized by statements and arguments about particular cases holding true only presumptively. *Phronesis* is practical prudence in dealing with particular individuals, specific problems, and the details of concrete cases or situations. To Aristotle it was clear that ethics was a matter of *phronesis*, not *episteme*. Moreover, it is obvious that *phronesis* in ethics implies casuistry, i.e. a case-by-case approach (see Jonsen, Toulmin 1988, pp. 23-46).

What is, more precisely, the relation between being a good person and performing right actions? Here, one might distinguish between three possible kinds of virtue ethics. According to the first, certain traits of character are a sufficient condition for performing right actions. If one has these traits, one will perform the right actions. According to the second, they are a necessary, but not sufficient, condition. If one does not have these traits, one cannot perform right actions, but having them is not a guarantee that one does. According to the third kind, such traits of character are neither necessary nor sufficient for performing right actions, but only contribute to performing such actions under certain circumstances. One may perform right actions even if one does not have these traits of character, but having them commonly contributes to performing right actions. This is likely how most virtue ethicists understand their position. However, it is not self-evident that this third form of virtue ethics should be classified as virtue ethics. Whether it should be classified as such is a matter of focus. One may agree that cultivating certain good traits of character is of vital importance in order to be able to perform right actions at a high frequency and nevertheless classify one's view as a kind of action-oriented ethics, not as a kind of virtue ethics. This is so because the primary focus is on actions, not on traits of character. The central question is: which actions are right and which are not?

In my opinion, imaginative casuistry should be classified as an action-oriented ethics. It stresses the importance of cultivating traits of character such as sensitivity, empathy, and imagination in order to be able to perform right actions. However, the primary focus is on action, not on traits of character. But let me stress that in general I am very positive to virtue ethics. As the cognitive scientist Francisco Varela has argued, it fits very well the empirical finding of cognitive science that most of the time in our everyday life we are involved in an immediate coping with what confronts us (our 'microworld'). We react spontaneously without reflective deliberation. This means that if we cultivate good traits of character, we will spontaneously do good things (Varela, 1999, pp. 3-6, 8-11, 23-24). Moreover, if we include *phronesis* as a virtue we can also account for those cases that require deliberation rather than spontaneous reaction.

Why then prefer imaginative casuistry to virtue ethics? The arguments in favor of imaginative casuistry have already been presented. My arguments against virtue ethics are as follows.

A classical objection is that even a good person might perform wrong actions, for instance, due to lack of information. Being a person with certain good traits of character is not in itself a guarantee for performing right actions, although the frequency of right actions is probably higher for such a person.

Another objection is that virtue ethics is not very helpful when we face new situations that cannot be subject to immediate coping. As Varela says (Varela, 1999, p. 18):

> It is at the moments of breakdown, that is, when we are *not* experts of our microworld anymore, that we deliberate and analyze, that we become like beginners seeking to feel at ease with the task at hand.

Even if we include *phronesis* as a virtue–which is precisely the ability of reflective deliberation in facing new and difficult situations–we must still actively use this practical wisdom in each particular case in order to solve the problems.

This objection is particularly serious when virtue ethics is applied to biomedical issues. Admittedly, certain virtues are truly needed in clinical practice. Physicians would do well to cultivate certain good traits of character. But virtue ethics does not constitute a very helpful framework in determining the right course of action regarding the high technology advances in biomedicine, for instance, in medical genetics. Much of this technology raises new ethical questions. We cannot cope with them spontaneously, neither can the mere reference to *phronesis* help us. We must carry out the ethical deliberation in practice. This means interpreting, applying, and ranking ethical principles in arguments for and against different alternatives of action, and this is exactly what action-oriented casuistry is all about. It is the practical conclusions with regard to particular cases that really matters, not whether those who carry out the ethical deliberation are virtuous.

In addition, the most important ethical questions raised by modern biomedicine concern professional policy. What is really needed in biomedical practice and research is a policy for all whether one has cultivated good traits of character or not.

Imaginative Casuistry and Responsibility

In the first part of this chapter, I discussed the concept of responsibility. I argued that this concept is metaphorical and can be used in many different ways. This means that the analysis was actually completely in line with the approach of cognitive semantics and imaginative casuistry.

I also stressed that it is possible to combine the different metaethical models of responsibility–agential, social and encounter models–with many different normative ethical views. This was true also of the modified social model that I proposed. I have now argued that imaginative casuistry is the most promising ethical position. Consequently, my overall suggestion is that the modified social model of moral responsibility be combined with imaginative casuistry.

Now, the question arises: what are the implications of imaginative casuistry for the view of responsibility? In one of its uses responsibility is understood as a virtue, i.e. a desirable trait of character to be cultivated to the level of excellence. This does not contradict imaginative casuistry. Developing the virtue of responsibility is compatible with the action-oriented imaginative casuistry. To be sure, due to its pluralistic character, imaginative casuistry recognizes many other virtues–as well as other values and principles–but the virtue of responsibility is particularly central. A responsible person uses his or her moral imagination in order to become aware of–listen to–the needs of others, and respond in a way that is sensitive to the context. A responsible person is not simply obeying certain strictly defined rules put forward in advance but makes context-sensitive judgments from case to case based on empathy and an envisionment of a plurality of alternatives and perspectives.

Thus, when imaginative casuistry is applied at the individual level, the virtue of responsibility is very important. At group and community levels and at the governmental level, however, it makes little sense to talk about responsibility in terms of a virtue. At group and community levels it is more appropriate to talk about good 'morals', i.e. good manners and customs. And at governmental level it is better to talk about policies and legislation with the public good as a goal. But imaginative casuistry is nevertheless useful at all these levels.

Applied to science this means, first, that individual scientists should cultivate the virtue of responsibility. Second, it means that the scientific community should cultivate good 'morals' in research, i.e. good manners and customs. Third, it means that the government should enact policies and legislation regarding research with the public good as a goal. At all these levels, imaginative casuistry should be the normative guide.

4. IMAGINATIVE CASUISTRY IN BIOETHICS

What are the implications of imaginative casuistry for bioethics? What is implied and what is not?

Imaginative casuistry implies that we reach a position in bioethics that is more psychologically realistic than other ethical views, including other versions of casuistry and principlism. The importance of metaphors and imagination is fully recognized.

Moreover, in imaginative casuistry one asks other questions than in traditional approaches. Instead of primarily asking questions about deduction from well-defined principles, one asks questions about metaphors. What metaphors determine health care? What metaphors are at the basis of thinking and action? In what direction should the metaphors of bioethics be extended?

This form of ethics also sheds new light on the nature of disagreements and agreements in bioethics. Disagreements often have to do with differences regarding in the direction and extent to which a metaphor should be extended. By going back to the original metaphor or by changing metaphor disagreements may be settled. One important goal of bioethics should be to reach an agreement on metaphors. The method to reach this goal is imagination and dialogue.

Finally, imaginative casuistry implies that the bioethicist should make more modest claims. Ethics loses its absolute character and its statements are viewed as tentative. Certainly, this is something that most ethicists would accept, but it acquires a stronger emphasis in imaginative casuistry.

Imaginative casuistry does not imply, however, that principles are without value. Ethical principles have their place in reflection and discussion, but one has to be aware that they do not have a strictly defined, literal meaning, and that it is not a matter of simply applying them to particular cases. The principles of bioethics are important rules of thumb based on collective experience. They express what is valid in prototypical cases. What should be valid in nonprototypical cases is a question of moral imagination.

Now, I would like to discuss the implications of imaginative casuistry for bioethics and research ethics more in detail. I shall focus on two kinds of concepts. First, I shall give some examples of metaphorical concepts framing situations in medical practice and research. Second, I shall analyze a number of central bioethical principles.

Metaphorical Framing of Medical Practice and Research

The importance of metaphors in the framing of situations (cases) is clearly seen in medical practice and research. One important metaphor is 'medicine is war'. This is a very common metaphor, both in actual medical practice and in discussions on metaphors in bioethics. Politicians, physicians and researchers talk about "a war on cancer" and "a war on AIDS." Presently, one often hears that medicine now "is fighting the war against cancer on the genetic battle-field." But there are many other examples: "The physician orders a *battery* of tests", "Let us treat *aggressively*", and "Good patients are those who *fight* and refuse to *give up*" (Childress, 1995, pp. 1765-1768; Sontag, 1990, pp. 65, 72, 99).

There are several advantages with this 'war' metaphor. It can be very effective in fund raising. The government might be more willing to contribute to the efforts of fighting certain diseases if these diseases are described as threats to the whole nation similar to enemies in war. Researchers might be influenced to focus their efforts on these kinds of diseases. Physicians might be inspired to do their utmost for helping their patients. Patients can in this way be exhorted to use all their inner strength in resisting their disease. A disadvantage with the 'war' metaphor, however, is that it might contribute to stigmatizing certain illnesses and, consequently, certain patients. This risk has been highlighted by Susan Sontag with regard to AIDS (Sontag, 1990, p. 99).

In these examples we clearly see that the source domain, war, structures the thinking and action in the target domain, medicine. This way of framing in medical practice might be a good thing in some cases, but a bad thing in others. The 'war' metaphor is in itself a mighty weapon that should be used with discretion.

Another metaphor that represents a completely different way of framing situations of medical practice is 'medicine is business'. In this metaphor, or rather complex of metaphors, disease or illness is seen as a demand of service, the patient as a consumer, the treatment as a service, and the doctor as a provider of service. In a private health care system this perspective naturally has its justified place, but it is obvious that if it is used as a metaphor, and as a dominating one, it might affect health care in a negative manner.

There is also a 'medicine is machine repair' metaphor. No one actually uses this metaphor in its pure form as a way of framing situations of medical practice. Nevertheless, in modified forms it has great influence in medicine and biomedical research, not least in gene therapy research. Within this metaphor, when the patient is healthy, he or she is imagined as a well functioning machine. Similarly, disease is understood as an objectively detectable machine failure. The doctor is seen as a machine repairer. If the doctor would coherently base his or her actions on this metaphor, this would have radical implications for the relations between doctor and patient. For instance, it would not be important for the doctor to obtain informed, voluntary consent from the patient. You do not obtain consent from a machine, or at least the metaphor suggests as much. The machine would simply be repaired. Now, of course, no one acts consequently on this metaphor. Most physicians and scientists do not accept this reductionistic view on human beings, but see patients as something more than machines.

An alternative metaphor is 'medicine is system maintenance'. This framing is inspired by systems theory. The body is seen as a feedback system. It is a whole with parts interacting in a balanced way. Disease is seen as a disturbance of the balance and the doctor as someone who aims to reestablishing the balance. An advantage of this metaphor is that it is not reductionistic, but holistic. Systems theory recognizes and tries to explain the emergence of new qualities on higher levels of organization, qualities such as those that we usually associate with personal human beings.[8] Moreover, since patients are seen as persons, this metaphor is also compatible with idea of obtaining voluntary, informed consent. However, it seems as if the metaphor is of limited value in everyday health care. It represents an important theoretical perspective, but the metaphor does not catch everything that people consider important in practical medicine.

Preferable from this practical perspective is the 'medicine is personal care' metaphor. 'Care' is here understood not only as 'competent treatment' which is the common meaning of the established expression 'health care', but also–and this is particularly important–as 'deep concern'. 'Care' in this latter sense is clearly metaphorical. The source domain is the deep concern people show each other within personal relationships, for instance within the family and among friends. This concern is transferred to the target domain of professional medical practice. Professional relations are seen as personal relations.

Furthermore, that care is 'personal' implies, among other things, that it is an attitude of a person, the doctor, towards another person, the patient, and, as indicated in relation to the 'system' metaphor, the conception of the patient as a person–in contrast to a machine–is compatible with the principle of obtaining voluntary, informed consent. Moreover, since the patient as a person has a subjective experience, the person's health and illness cannot be reduced to an objective dimension. A subjective element is included. The patient experiences himself or herself subjectively as healthy or ill. Consequently, if we accept that the doctor should treat the patient as well as show deep concern for him or her, this implies that the doctor also should take the patient's subjective experiences into consideration. The doctor cannot, as in the 'machine' metaphor, focus exclusively on the objective aspects. This might also have implications for clinical research; perhaps the subjective experiences of the patients should be studied more thoroughly.

This brief presentation shows the importance of metaphors in framing medical practice and research. Different metaphors have different implications for thinking and acting. Consequently, we should be very careful which we choose.

---

[8] For a more elaborate discussion of the implications of systems theory for the view on human beings, see Nordgren, 1994, pp. 56-57, 98-101.

The Metaphorical Character and Prototype Structure of Central Bioethical Principles

The modern discussion in bioethics and research ethics has focused to a large extent on the following basic principles: autonomy, utility, justice, and integrity. I accept these as rules of thumb summarizing the collective moral experiences of the Western society including the scientific community (see p. 35). Below, I will try to establish that they are all metaphorical and exhibit prototype structure, and that, consequently, bioethics is not simply a matter of applying these principles to particular cases. Generally, I will not take a stand regarding specific interpretations of the principles. In line with the proposed casuistic approach, different interpretations may be appropriate in different cases. Only occasionally will I take a stand.

The Metaphor of Autonomy

Respect for autonomy is a central principle of modern bioethics and research ethics. But what is autonomy? What is it that should be respected? I would like to show that autonomy is basically a metaphor and that this metaphor has been extended in several different ways.

Autonomy is fundamentally a metaphor from the political domain. Originally the term referred to the self-government and independence characterizing the ancient Greek city-states. The self-determination was practiced by independent states that had the ability to institute laws for themselves (Gr. *auto* = self, *nomos* = law). This originally political term has also been metaphorically applied to morality as early as in antiquity (Pellegrino, 1990, p. 4; Bischofberger, 1990, p. 48). However, while in the political context it is a matter of enacting laws within the state together, in the moral context it is a matter of enacting laws for oneself as an individual. In this, the metaphorical character of the moral application of the term 'autonomy' becomes obvious. In a metaphorical, but not literal, sense one can institute laws for oneself.

In the modern moral context, the metaphor of autonomy has been extended in several different ways, and these extensions imply different interpretations of the principle of respect for autonomy. First, autonomy has been seen as a radical ability that is metaphysically grounded. We are autonomous in the sense that we have a free will that stands outside the causal network of nature and history. Autonomy is an ability that we can and ought to practice, even if we often fail to do so. Every person is autonomous in this radical sense. To respect the autonomy of a person is to let him or her use this ability. In the medical context, this implies that the doctor gives the patient full information about diagnosis, treatment and consequences of treatment. Personally I find it difficult to accept the idea of a free will that stands outside the causal network, although this is not the place to present a lengthy argument.

Moreover, autonomy is sometimes understood as an empirically grounded ability within the causal network of nature and history. We have a free will that is biologically and socially situated. Autonomy implies intentionality and understanding, as well as the absence of a single controlling factor determining the act. But because understanding and determination are matters of degree, there are degrees of autonomy. People can be more or less autonomous. To respect someone's autonomy is to stimulate him or her to decide what to do. This implies that the doctor organizes the clinical environment in such a way that it encourages the patient to choose actively among different options. The patient is dependent on the doctor but must be stimulated to make his or her own decision about what to do.

Autonomy is also understood as something ascribed to people, for instance patients, when their choices are respected. Every individual should be respected independently of the values he or she chooses. To respect a person is to respect his or her choices. The focus is not, as in the other two metaphorical extensions, on the ability of autonomy but on the respect for the choices whatever they are, at least to the extent that they do not lead to acts that harm other people. In health care, there are always some restrictions on autonomy in this sense. Respect for autonomy is interpreted more or less radically. For instance, there are more moderate, as well as more radical, views on the reproductive autonomy of prospective parents. Sometimes autonomy in this sense is stated as a right to choose. In this case, the metaphor of autonomy is combined with a financial metaphor of credit.

Finally, autonomy is something ascribed to people when they are choosing how to act on a basis of a will that is rational, in accordance with a universal moral law, and pure in the sense that it is not contaminated by emotions or personal preferences. This is the Kantian interpretation of the concept of autonomy (see Lakoff, Johnson, 1999, pp. 415-439). Autonomy is to Kant a normative concept, and only now and then are we humans able to live up to its standard. Commonly, we are determined not by our free will but by our emotions, inclinations and preferences. Furthermore, the principle of respect for autonomy is, in Kantian thought, based on the categorical imperative, i.e. that we should treat other persons always as ends in themselves and never as means only. This implies that we should respect other people's decisions even if they are not autonomous in the full Kantian sense. As was the case with the metaphysical understanding of autonomy mentioned above, I find it difficult to accept the Kantian view. It seems to be at odds with the empirical findings of cognitive semantics.[9]

In sum, there are many different interpretations of the concept of autonomy. They all represent different extensions of the original metaphor. In coming chapters, the principle of respect for autonomy will be very central, not least in our discussion

---

[9] See the discussion of Kantianism on pp. 25-28.

of reproductive autonomy. In these discussions, it is important to be aware of the plurality of interpretations of the concept.

The Cost-Benefit Metaphor

The principle of utility is another important principle of bioethics. It is based on a 'commercial transaction' metaphor and a 'wealth' metaphor. Utility is what leads to, for instance, well-being, pleasure, happiness or preference satisfaction, which are understood in terms of wealth, i.e., as access to valuable goods.

An application of this 'commercial transaction' metaphor is the distinction between 'costs' and 'benefits'. The goal of business is to maximize profits and minimize costs. In the moral context, one reasons in a similar way. It is a matter of maximizing good consequences and minimizing harmful ones. This view is manifested in bioethics in utilitarian and other interpretations of the principles of beneficence and nonmaleficence.

In the medical context, an important problem is: utility for whom? Traditionally, the object of concern has been the patient at hand. However, today many ethicists would take into consideration future patients and the society as a whole as well. This is also a problem of metaphorical extension and moral imagination. Should utility be seen as utility for the patient at hand or as social utility?

Justice: A Metaphor of Balancing

We now turn to the concept of justice. One basic metaphor used to understand this concept is a balance in which equally heavy weights lie in each of the scales. The concept of justice has been extended in at least two different ways. In one extension justice is understood retributively, i.e., as retribution of evil and good deeds, and in another extension distributively, i.e., as distribution of the good in society. Distributive justice is what is commonly intended in the modern debate in bioethics and is the concept I shall focus on here.

Different principles of justice have been suggested. Each represents different extensions of metaphors, 'balance' metaphors and others. Below follows a presentation of some of these and an explication of the metaphorical elements.

First, we have the principle of equality. This principle is based on a clear 'balance' metaphor in which equally heavy weights lie in each of the scales. At least two different metaphorical extensions have been suggested. According to one, equality is achieved when everyone involved receives equal amounts of resources. According to the other, it is achieved when everyone receives equal opportunities to experience health and welfare. This may require that they receive different amounts of resources.

Another principle of justice focuses on merit or desert. This principle presupposes a metaphor of proportion, i.e., one starts from certain measureable properties or performances, and maintains that the individuals involved should be given the goods in proportion to these measurable properties. In health care priority-setting, these properties might be, for example, age, whether low or high, or social utility.

Justice may also be articulated in terms of a contract between people. This view is directly linked to a 'commercial transaction'. According to this view, people who contribute to society by paying taxes have the right to use the social resources, for instance health care, when they need them. More precisely, it is a metaphor of 'credit'.

Justice is also interpreted in terms of needs. The principle of need means that goods should be distributed in such a way that patients with the greatest needs get the most. In this case we have an economic metaphor that emphasizes the importance of taking into consideration the consumers' needs of different goods, combined with a 'balance' metaphor according to which everyone should be given the same opportunities to experience welfare.

A quite different conception of justice focuses on rights. Here we find at least two different extensions, both based on a 'commercial transaction' metaphor. Rights are interpreted in terms of credit. One extension focuses on the rights of patients, and the other on the rights of property owners. In the latter case, it is argued that it is unfair for the state to take with force from those who have righteously acquired their property and give to other people, for instance patients, with greater needs. The right to property is a human right and a fair society should respect this right.

Finally, there is a utilitarian notion of justice. According to this view, justice is utility. A distribution in health care is just when the resources are allocated where they do the most good. The metaphor is purely economic.

All these different metaphorical extensions may be relevant in the medical context. The fundamental problem is: in what directions and how far should the metaphor of justice be extended?

The Metaphor of Integrity

The word 'integrity' originates from the Latin term '*integritas*' (and '*integer*') which refers to wholeness, completeness, untouchedness, unimpaired unity, coherence, and harmony. The Romans considered an integrated system to be a whole where the parts contribute to the harmony, i.e., a well coordinated system. And regarding human beings, they understood integrity as an ethical concept. A human being was metaphorically viewed as a harmonious system (Pellegrino, 1990, p. 10; Bischofberger, 1990, pp. 47-48). According to this view, an integrated person is a person with coherent character and who exhibits coherent behavior. An integrated person is a well coordinated person. Thus, integrity is seen as a moral virtue (Pellegrino, 1990, p. 10).

But this is only one way of doing the metaphorical extension. The concept has also been applied to persons who do not exhibit a coherent character. It can refer to the human dignity shared by every person (Pellegrino, 1990, p. 10; Bischofberger, 1990, p. 48). As units or wholes, persons have a value in themselves independent of all other functions or properties of character. While this use of the term 'integrity' is uncommon in the British and American context, it is prominent in, for instance, the German ('integrität') and Scandinavian (the Swedish word 'integritet') context. A clear example can be seen in a report on genetic engineering to the Council of Europe from 1982 (Council of Europe, 1982):

> The principle on which all existing human rights are based is the idea of man's integrity... A person's chromosomes, genes, etc., must clearly be a part of what is covered by this integrity just like his spirit, flesh, and blood.

Respect for integrity in the sense of human dignity is, of course, often interpreted in a Kantian way as a prescription of treating other persons never as means only but always as ends in themselves. Moreover, integrity in this sense may constitute the basis for the notion of human rights.

A third kind of metaphorical extension is related to respecting and protecting the private sphere of persons. This sphere is a whole that should not be violated, i.e. broken up. Here, it is common to make a distinction between physical integrity, i.e. the physical sphere, and mental integrity, i.e. the mental sphere. In the former case, it is a person's body that should be treated with respect, in the latter it is a person's feelings, beliefs and so forth. The distinction is not strict, of course. Disrespect for a person's body commonly hurts his or her feelings. However, many people believe that even the dead body has integrity.

In the medical context, all three kinds of extension are seen. Sometimes the focus is on integrity as a virtue of the medical staff, sometimes on the principle of respect for the integrity of the patients in terms of human dignity, sometimes on the principle of respect for the integrity of the patients in terms of privacy and confidentiality of medical information.

Integrity is intimately linked to the concept of a person. However, the question is how far this concept should be extended. Who has integrity? As mentioned above, white, adult, heterosexual and healthy males have often represented the prototypical cases, while females, children, fetuses, nonwhite people, homosexuals and the mentally retarded have been understood as nonprototypes. Should all humans be viewed as persons and should animals other than humans possibly be viewed as persons as well?

The concept of integrity has also been applied to entities other than persons. Some people talk about the 'integrity of Nature'. Some theologians talk about the 'integrity of Creation'. Nature is understood as a whole that ought not to be broken up. The 'natural order' should be respected. The concept of integrity has also been extended to the genome, as indicated by the quotation above. One talks about 'genetic integrity'. The genome is a whole that should not be broken up. The overall question is: what is the scope of application of the concept of integrity?

Conclusion: Different Interpretations, Applications, and Rankings of Ethical Principles

This analysis of central ethical principles in bioethics and research ethics indicates that although there is a general consensus at an abstract level that principles such as autonomy, utility, justice and integrity are important, there is great disagreement about their interpretation and scope of application. In addition, as I have stressed earlier in the chapter, the principles may be ranked differently when they conflict. All this is what could be expected from the perspectives of cognitive semantics, moral tinkering, and imaginative casuistry. It is vital that this be recognized in all discussions in bioethics and research ethics.

CHAPTER TWO

MORAL RESPONSIBILITY IN SCIENCE

1. RESEARCH ETHICS

What does moral responsibility in science mean? Are experienced scientists more responsible than junior scientists? What responsibility lies with the scientific community? For what are scientists responsible? Do scientists have a responsibility for the applications of their findings? To whom are scientists responsible? Do scientists have a responsibility to future generations and to animals used in research? Is there a 'real' moral responsibility in distinction to the responsibility ascribed to scientists by different social groups? What light does the proposed modified social model of moral responsibility shed on the social activity of science? How can the approach of imaginative casuistry be used by scientists? These are some of the questions to be discussed in this chapter.

There are two different traditions in research ethics. The most influential has focused primarily on 'internal' issues such as scientific misconduct and the treatment of human subjects. A clear example is the *Helsinki Declaration.*[1] To the extent that 'external' issues such as the social responsibility of the scientific community are explicitly addressed, they are stated only in general terms: 'science should aim at the public good.' Often, it is implicitly assumed that science in and of itself is for the good of society.[2] However, there is another tradition in research ethics focusing more directly on science as a part of society and on social responsibility. Often this kind of research ethics has taken a partly critical view of the role of science in society: 'science is sometimes used to legitimize established class and gender structures.'[3] Examples can be found not only in socialist and feminist writings, but also in liberal ones.

The time is ripe for connecting these two traditions in research ethics. Actually, the responsibility of scientists doing human genetics research raises both 'internal' and 'external' questions. The internal issues include the question of maintaining

---

[1] See, for instance, Annas, Grodin, 1992, p. 331-336. See also URL: www.wma.net.

[2] In this tradition can also be included the 'CUDOS norms' articulated by Robert Merton (1973 (1942)): Coummunism, Universalism, Disinterestedness, and Organized Scepticism.

[3] This tradition is not as homogeneous as the dominating one. It includes, for example, socialist and feminist critiques of science. Two books clearly illustrating this tradition are Rose, Lewontin, Kamin, 1984, and Hubbard, Wald, 1993.

high scientific quality and avoiding ideological bias and overinterpretation in genetics research regarding issues that are politically heavily charged, and the question of the responsibility to human subjects and animals participating in genetics research. The external issues include the question of what should constitute socially and morally responsible choices of subjects (topics) and the question of the responsibility of scientists for the social uses of their findings. In the end, however, the distinction between internal and external issues can hardly be maintained. They are intimately connected.

Different Kinds of Responsibility in Science

There are many different kinds of responsibility in human genetics research. There is a purely scientific responsibility, which can be understood as a responsibility to observe certain minimum quality standards in all aspects of scientific work. Another kind of responsibility is legal. Like other citizens, scientists are obliged to follow the general laws of the country in which they work. There might also be other laws exclusively regulating scientific work, although these may not be so common. We may also talk about an economic responsibility–research facilities often cost a great deal of money. With this comes an obligation for responsible management. There is also a social responsibility: scientists have a responsibility to society for their professional activities. This is particularly true when their activities are paid for by taxes. Finally, scientists have a moral responsibility. This kind of responsibility is commonly understood as more fundamental than the other kinds. However, according to the proposed modified social model of moral responsibility, I do not make a distinction between socially attributed moral responsibility and a 'real' moral responsibility. Social practices of blame and praise is all we have. There is no 'real' moral responsibility beyond the one ascribed by the community. Moreover, it should be noted that moral responsibility in science is not only a matter of personal ethics but also of governmental and professional policy.

I shall focus primarily on moral responsibility. However, it is obvious that it interacts with other kinds of responsibility. Morality may, for instance, support the demand for high scientific quality. It may also back up legal regulations as well as demand criticism of existing laws.

An Analytical Model: Responsibility 'of', 'for', and 'to'

In order to give the analysis of moral responsibility in science a useful structure, I shall take as a starting point the distinctions made in the previous chapter between the aspects of responsibility *to*, responsibility *for*, and responsibility *of*. As I use the word 'for', it only applies to actions and their consequences, not to people, animals, etc. 'To', on the other hand, applies to people, animals, etc. The aspect of responsibility 'of' refers to those who are responsible such as individuals and groups. When applied to science, the three aspects can be caught in the following three questions:

- Who is responsible in science? (responsibility *of*)
- For what are scientists responsible? (responsibility *for*)
- To whom and to what are scientists responsible? (responsibility *to*)

With regard to the first aspect we are talking about the responsibility of scientists, but do all scientists have the same responsibility? For instance, is there a difference in responsibility between researchers high in the scientific hierarchy and junior researchers? Is there a difference in responsibility between scientists doing basic research and those doing applied research? Is there a difference in responsibility between scientists working in universities and those working in private companies?

The second aspect is responsibility *for*. Scientists are responsible for their actions and their consequences. Here I find it useful to focus on the different stages of research and its application: choice of subject (topic), the research process in the narrow sense (empirical and theoretical work), publication, and application (use). It is uncontroversial to argue that scientists are responsible for their choice of subject, the research process in the narrow sense and the publication of their results, although there are many specific ethical problems raised by each of these stages. However, the view that they are responsible for the application of their findings is more controversial. Do scientists truly have responsibility for applications of their results that other people make?

The third aspect–responsibility *to*–is also important to discuss in relation science. Scientific research affects many different parties, e.g. human subjects, funding agencies, and the general public. Conversely, these parties affect the scientists. Exactly which parties are the scientists responsible to? How should they respond? What do they owe to different parties? And to what extent are different kinds of scientists responsible to different parties?

Below I shall apply this analytical model to science in general and to human genetics in particular.

Domain, Content, and Form of Responsibility

I call the linkage of a stage of research (*for* which scientists are responsible) with a party (*to* whom scientists are responsible) a 'domain of responsibility'. An example of such a domain of responsibility is the responsibility to human participants in the research process. Another example is the responsibility to the general public to the extent that it is affected by the applications (uses) of research.

It is important to distinguish the domain of responsibility from the content of responsibility. By the 'content of responsibility' I mean views concerning what scientists should do. Different ethical positions might imply diverging suggestions regarding this normative content. Two people might agree about a certain domain of responsibility but disagree about the specific content of the responsibility. For instance, they might agree that scientists have a responsibility to human subjects or animals used in the research process but disagree about how the human subjects or animals should be treated. Furthermore, to take an example from human genetics, two persons might agree that geneticists should assume responsibility for the applications of genome research but disagree about what this means more specifically. For instance, many geneticists in the beginning of this century obviously saw themselves as taking responsibility by promoting eugenics. However, critical voices today say that eugenics is something that should be strongly rejected as a part of responsibility.

Moreover, to be distinguished from the domain of responsibility as well as the content of responsibility is the 'form of responsibility'. By this I mean the level at which the normative content should be implemented in action and the means of doing this. For instance, avoidance of scientific misconduct can be approached at the level of government, at the level of the scientific community, and at the individual level. The means of achieving this could be legislation by the government, cultivation of good 'morals', i.e. good manners and customs, within the scientific community, or development of personal virtues of individual scientists. There can be disagreement about level of implementation and means among those who agree about the content of responsibility.

2. RESPONSIBILITY 'OF'

I shall begin with the issue of moral 'responsibility of' scientists. I view this moral responsibility from a social perspective: the moral responsibility of the scientists is a matter of social practice and social role. This raises a number of fundamental questions. How should the interaction between science, technology and society be understood? Is the development of science and technology and its application in society a matter of choice, and if so, to what extent? Are there important differences in responsibility between scientists doing basic research and those doing applied research? What responsibility lies with the individual scientists, the research groups, and the scientific community at large? What implications does the distinction between 'holding responsible' and 'assuming responsibility' have in this context? Are there differences in responsibility between researchers at different levels in the scientific hierarchy? What different roles may individuals assume within the scientific community and what responsibilities do these roles imply? What is the relation between political control of research and moral responsibility? Let us take a closer look at these questions and their different answers.

Science, Technology, Society

It is not too uncommon a view that technological development has a momentum of its own, which makes it virtually impossible to affect. Sometimes this argument is also applied to the development of gene technology. Knowledge of the human genome–it is maintained–will drive society in a certain direction whether we want it or not. Certain applications will emerge with necessity and it will be impossible to stop them. In this study, I hold that this 'technological determinism' is inadequate. I hold that decisions made by individuals and small groups can affect the course of technological development, at least to some extent. Moreover, I maintain that this is true of the development of society as well. These assumptions are practically necessary. Without them there is no reason to try to affect future developments and without them it is also meaningless to talk about moral responsibility. However, this 'voluntarism' has certain limits. Our expectations have to be realistic. Our ability to control the future is limited. The extent to which we can affect future developments can only be discovered in practice.

On the basis of these assumptions, the consequences–or applications–of science in society become not just a matter of anticipation but also, at least to some extent, a matter of choice. When we talk about, for instance, the place of genetics in the future of society, we do not talk about something that is predetermined but about something that we, at least to some extent, may choose.

These assumptions cannot be demonstrated beyond doubt. However, in my opinion, some good reasons can be offered and I will try to do so below.

A Dynamic Systems Approach Combined with a Limited Voluntarism

What is the relation between scientific, technological and social development? What driving forces are primary? Can decisions made by individuals and groups determine the course of development? These questions are, of course, very difficult to answer. One reason is that all the concepts–science, technology, society–are very difficult to define. For my purposes, however, it does not seem necessary to give any strict definitions of science and technology. In this book, I focus primarily on human genetics and gene technology, and these phenomena fall, to be sure, under any concept of science and technology. Society, on the other hand, is a more unclear concept, and I will come back to it below. Let us first take a brief look at three major models of the driving forces in scientific, technological and social development (cf. Barbour, 1992, pp. 20-21).

According to the first view, science is the primary cause. Technology is an application of science but affects the scientific development only to a limited extent. On the other hand, it does affect social development. Society in its turn hardly affects science and technology. This position can be called 'the primacy of science' view. It is a linear view where science generates technology which in its turn affects society. Examples from the history of science and technology that could be put forward in favor of this view is the research in the fields of electricity and chemistry during the second half of the 19th century which generated powerful technological innovations and radical social change.

Another position is 'the primacy of technology' view. Here technology is the primary cause of change. It affects both science and society but these affect technology only to a limited extent. Technology drives science and society uncompromisingly ahead. It is a matter of technological determinism. In order to explain human history, the focus must be on technology. Marxism has sometimes been interpreted in this way. One example of the primacy of technology is the steam-engine which generated thermodynamics and was also of immense significance for the development of early industrialism.

A third position is 'contextual interactionism'. This view is contextual, because it stresses that science and technology always belong to a particular social context. It is interactionist, because society, science and technology are understood as factors which all interact and influence each other's development. It is stressed that not only can science and technology be driving forces in social development but the social context can affect scientific and technological development as well. An example of the influence on scientific development is when politicians allocate more economic resources to certain kinds of research than to others. An example of the influence of society on technological development is when politicians put certain environmental restrictions on the technology to be used in society such as catalysts on cars in order to reduce pollution.

Contextual interactionism seems to be the only acceptable model. The examples given show that we have to accept the complexity in the scientific, technological and social development. However, a weakness of this kind of theory, as it has been presented so far, concerns the unclear concept of society. What is society? What is social context? To be sure, 'society' and 'social context' are very complex phenomena. Society was a Roman invention. A *societas* was a grouping for a specific purpose, much like the word 'association' in modern English. 'Society' can be defined as a set of overlapping communities and organizations that shapes the individual self and provides the context in which the individual accepts responsibility for certain actions (cf. Moran, 1996, pp. 119, 122). In this sense society can be described as a continuum from very loosely associated groups of people (communities) to business corporations (organizations). Two kinds of institutions (corporations, organizations) of particular importance for science are the government and private companies. Moreover, science as an organization and a community of people is in itself also a part of society. Thus, when we talk about science as interacting with society, we use the word 'society' as shorthand for nonscientific institutions such as the government and private companies, and nonscientific communities such as interest groups and families. The issue is actually even more complex, since some scientific research is done within universities and some in private companies.

Another weakness of contextual interactionism, as it has been presented here, is that the emphasis on social context might lead to an underestimation of the roles of individuals and small groups. Certainly, the view is congruent with the experience that individuals and small groups can seldom affect history in a decisive way. The social context is crucial. However, it also seems obvious that individuals and small groups can sometimes fundamentally change the course of events.

With this in mind, I would like to suggest a 'dynamic systems approach' to the problem at hand.[4] From this perspective, science, technology, and society can be described as an evolutionary, complex, nonlinear system. Note, however, that this description, at least for now, is only a metaphor. No one has as yet found a way of describing such a complex system mathematically. Within this metaphor, individuals commonly cannot affect development to a significant extent. However, at certain critical points in the system's development this might be possible. Minimal differences such as the actions or views of one individual instead of another, or one small group instead of another, can have vast consequences. Individuals or small groups can sometimes marshal support of others for their views.

However, a recognition of the significance of the individual at critical points in the development of complex, nonlinear systems is not enough to support the view that decisions made by individuals and small groups can determine the course of scientific, technological and social development. The dynamic systems approach does not in itself support the assumption that decisions of individuals might be a driving force. However, it is compatible with such an assumption. A limited scientific, technological, and social voluntarism is theoretically possible. It is also practically necessary; otherwise there is no reason to try to affect the development, and otherwise it would not be meaningful to talk about moral responsibility in science.

Another aspect of great importance to the issue of responsibility has to do with the relation between the individual and society. We have already met this problem in the previous chapter. According to social atomism, the individual exists prior to and independently of society. Society is viewed as a contract between individuals. Social holism, on the other hand, implies that the individual is born into an already existing society, which in its turn constitutes (shapes) the individual self. I have suggested a modified social holism that recognizes the role of society but nevertheless stresses the plurality of social (and biological) influences in the constitution of the individual self. Each individual is a unique combination of such a plurality of influences. The individual is more autonomous than traditional social holism would have it (see pp. 5-7). When applied to science, this view opens the possibility of the individual scientist or small groups of scientists being at odds with society or the scientific community at large. This means that the individual or group might assume other responsibilities than society or the scientific community demand. In this way, they

---

[4] For an extensive analysis and discussion of some of the implications of this kind of approach, see Nordgren, 1994, pp. 56-66, 119-139. An important part is the 'chaos theory'.

can criticize, for instance, injustice in society or the scientific community, and work for the elimination of social roles and traditions that discriminate.

In sum, I have argued for three fundamental assumptions of great importance, namely a qualified contextual interactionism concerning the driving forces of scientific, technological and social development, a limited voluntarism concerning the scientific, technological and social development, and a modified social holism concerning the relation between the individual scientist on the one hand, and society and the scientific community on the other. All three assumptions are relevant when we discuss the moral responsibility of scientists in general and geneticists in particular. To put it briefly, the development of technology, including gene technology, and its application in society is a matter of choice, at least to some extent.

Intended and Unintended Consequences

However, even if we can influence the technological development, there is always a possibility of unintended consequences, in contrast to intended ones. Intentions matter, but they are not wholly determinate. In addition, there might also occur unanticipated consequences in contrast to anticipated ones, and in both cases we have to distinguish between short term and long term consequences. Thus, in any discussion of the consequences–and applications–of science in society, we have to be aware of certain distinctions related to the concepts of intention, anticipation, and temporality. We have to make distinctions between intended and unintended consequences, anticipated and unanticipated consequences, and short term and long term consequences. Intended, anticipated, short term consequences of science is one thing; unintended, unanticipated, long term consequences, quite another. In addition, the ethical acceptability of the consequences might differ. Some applications are ethically acceptable and others ethically unacceptable. The scientists' responsibility is to promote good consequences and avoid bad ones. To be able to do this they have to anticipate the consequences. This is sometimes very difficult, and at other times quite easy. With regard to human genetics, it is often rather easy to anticipate at least some of the consequences. This is so, because the time lag is often short between scientific finding and application, and possible applications are often suggested in grant applications.

However, there is also another aspect that must be recognized in this context. In the promotion of a scientific project, there might be embedded objectives that are different from the overt objectives. The latter are those motives used to 'sell' the project. The former are other motives, known only by the project leaders. This distinction between overt and embedded objectives may be of some importance for the understanding and evaluation of the Human Genome Project. Alexander Rosenberg suspects that the real motive behind the sequencing project was a desire for technological spin-offs: research methods, computer programs, automatic systems for sequencing, etc. A complete sequence, including the sequence of the 'junk-DNA', would be of doubtful scientific interest (Rosenberg, 1996, p. 12). I do not agree on the latter but technological spin-offs were certainly a very important objective, although it was not stressed publicly.

Basic Science vs. Applied Science

It is important to understand the differences between basic science and applied science. One reason is that the distinction has relevance for the question of responsibility in science.

Basic and applied research can be distinguished by focusing on a difference in motive. The goal of basic science is to expand human knowledge. This is done for the sake of this knowledge itself, not for possible applications of it. It is a matter of discovering and understanding some aspect of the world, not of inventing something. Applied science, on the other hand, aims at conserving, changing or improving the human condition. It is designed to solve practical problems.

One may also locate the difference between basic and applied science in the activities as such and the products of the activities. Basic science produces theoretical and empirical knowledge. Applied science, on the other hand, produces technology or practical knowledge.

A third way of describing the difference between basic and applied science is in terms of different 'internal goods' (cf. MacIntyre, 1985, p. 187). Internal goods are values that are constitutive for a practice. Without these internal goods the practice would not be what it is. An internal good of basic science, viewed as a practice, is to understand the world. An internal good of applied science could be to change the world (or keep it unchanged). It is important to observe that this description is a matter of idealist abstraction. Actually, it is too idealist to be useful in the analysis of real science. A focus on motives and activities seems more promising. However, in discussions of choices of subject in research it can be of some use. The internal goods of a research project in basic research should be distinguished from external ones. Take the Human Genome Project. The main internal good of this project is to obtain genetic knowledge about humans (and certain model organisms). An external good could be to strengthen the nation economically and technologically by certain useful applications.

The distinction between basic and applied science is not always clear. A gray zone may be found in which we find research that is partially basic, partially applied. One way of describing the relation between basic and applied science is to view them as end points of a continuum. Specific research projects can be more or less basic or applied. For instance, the research on fusion energy may be regarded as both basic and applied. There is a clearly applied goal to this research, but there are so many technical obstacles that it may take another 30 to 50 years before we see a fusion reactor in use.

In biomedical science, the traditional distinction is between therapeutic and nontherapeutic research. This is clearly expressed in the *Helsinki Declaration* (1964, 1975, 2000).[5] In this ethical code, it is also obvious that the distinction has certain ethical implications. Different kinds of biomedical research imply, at least to some extent, different moral responsibilities. In therapeutic research, for instance, the researchers have special duties to patients.

How should human genetics research be understood? From the perspective of activity, it seems quite clear that genome research in the narrow sense is purely basic and nontherapeutic. It is a matter of mapping and sequencing the human genome and analyzing the functions of genes. However, the resulting basic knowledge is often rapidly put into practical use. It forms the basis for medical applications such as genetic testing and genetically designed drugs. The time lag between basic research and attempts to apply the findings is often short, and the applications are often the foreseen and intended results of the mapping and the sequencing. From the very beginning such applications are proclaimed to be the expected result of genome research. In addition, possible medical applications are often referred to as justification of the basic research projects in fund raising. Consequently, a focus on motives makes it clear that a sharp distinction between basic and applied genome research does not seem adequate in this case. The motives are mixed. For the same reasons, this is also true of the distinction between nontherapeutic and therapeutic research.

Different kinds of research may have different implications for the question of responsibility. Below I will argue that scientists doing strictly applied research have full responsibility for the applications they develop. As people in general, they are responsible for their direct actions. I will also argue that scientists doing purely basic research where no applications can reasonably be envisioned have no responsibility for possible future applications on the basis of their findings. As no applications are foreseeable or even conceivable they cannot be held responsible if such applications nevertheless are made in the future. However, to the extent that the applications are foreseen and intended, and are developed only a short time after the basic findings, basic researchers have, at least indirectly, a responsibility for them. This is obviously the case in human genome research.

Research at Universities and in Private Companies

Whether the research is conducted by public or private funding may also be relevant for the issue of moral responsibility. Many scientists work at universities, others in private companies. Among those who work at universities, some have public funding and others have private funding. Basic research is commonly publicly funded, while applied research is commonly privately funded. We see this also in

[5] See, for instance, Annas, Grodin, 1992, p. 331-336. See also URL: www.wma.net.

genetics. Applied genetics research is commonly conducted by scientists working in private companies. Basic genetics research, on the other hand, is commonly paid for by public funding or by foundations such as the Wellcome Trust. It is important to note, however, that in genetics a substantial part of the basic research is carried out by private companies. Prominent examples in gene sequencing are Celera Genomics and Incyte Pharmaceuticals.

Scientists who work in a private company or are privately funded may face a conflict of responsibility. The primary goal of private companies is to make good profit, not the public good. Certainly, making a good profit is often compatible with the public good, but there might occur conflicts of interest and such conflicts could imply conflicts of responsibility. The scientists have a responsibility to their company as well as to society at large.

Individual and Shared Responsibility

In old times scientists often worked alone. Today, however, it is quite the opposite. Modern science is to a large extent team-work. Within the university setting we find scientists forming research teams and belonging to departments. Moreover, these teams and departments often cooperate with teams and departments at other universities, both nationally and internationally. One clear example of international cooperation is the gene mapping and sequencing coordinated by the HUGO.

It is obvious that the collective character of the scientific enterprise has implications for the question of responsibility. When analyzing moral responsibility in science, one has to distinguish between the responsibility of

- individual scientists in different positions within the scientific hierarchy, at universities or in private companies,
- research teams at universities or in private companies, and institutions like departments, universities, etc., and
- the scientific community, within a specific discipline, within a specific set of disciplines, or at large, nationally or internationally.

The responsibilities may vary between these 'parties' as may the power to control.

How should responsibility be understood, given the collective character of scientific research? It is obvious that at the bottom line there is always an individual responsibility. The individual scientist has a responsibility for his or her own actions. In addition, this is a responsibility to the research team, the scientific community, the general public, etc.

But there is also a shared responsibility, because scientists work together. In fact, shared responsibility, rather than a strictly individual responsibility, is a distinctive feature of present day science. Within the shared responsibility, however, the responsibility of individual scientists may vary depending on the individual's place in the scientific hierarchy. Experienced and powerful researchers high in the hierarchy have more responsibility than doctoral students who play a limited role in the scientific community. Shared responsibility in this sense is distributed, but it is possible to talk about a nondistributed, shared responsibility as well. In this case, each member has the same responsibility.

Scientists may have shared responsibility not only for actions and omissions, but also for social roles, traits of character, and attitudes developed within the group. The shared responsibility is a responsibility to the individual scientists within the group but also to other teams, etc. Within research teams and the scientific community at large certain manners and customs are cultivated. It is obvious that the individual scientists have a shared responsibility for this 'climate' or 'morals' of research.

As an example, let us take the Human Genome Project. All individual geneticists taking part in the HGP have some responsibility for this enterprise and its consequences. How much the individual geneticists are responsible varies depending on the individual's role in the project.

What is most important, the responsibility of the scientific community or that of the individual scientists? From one perspective, the responsibility of the scientific community is the most important, because the individual scientist has a very limited control over the scientific development and the applications of scientific results. If one argues for increasing the scientific control, the shared responsibility of the scientific community is the proper starting point. The main question becomes how to organize the scientific enterprise in order to achieve this goal. However, whatever control the scientific community has, the question of the moral responsibility of individual scientists is still important.

A final remark: below when I talk about the moral responsibility of scientists without specification, I refer to the shared responsibility of the community of scientists.

'Holding Responsible' and 'Assuming Responsibility'

So far I have discussed 'being responsible' and 'having responsibility' without any explicit reference to the aspects of 'holding responsible' and 'assuming responsibility'. However, these latter aspects are keys to the former. To be responsible or to have responsibility is a matter of being held responsible by others or assuming responsibility by oneself. This is a central idea in the proposed modified social model of moral responsibility. Moreover, the distinction between holding responsible and assuming responsibility is essential when it comes to responsibility in science.

Science is a social activity. This makes it natural that people and institutions outside the scientific community make moral judgments regarding scientific activities. Nonscientists hold scientists responsible for research and its consequences. Scientists, on the other hand, make moral judgments about their own enterprise. They take responsibility for their own research. It is obvious that the judgments of scientists and nonscientists might conflict.

This social ascription and assumption of responsibility is all there is to moral responsibility in science. In line with my argument in the previous chapter, to talk about a 'real' moral responsibility beyond the moral responsibility ascribed to scientists by other social groups and beyond the moral responsibility assumed by the scientists themselves is theoretically open, practically meaningless and practically uninteresting. The issue of moral responsibility in science is an issue of social practice, and different social groups and individuals within and outside the scientific community may be of different opinion. This means that the issue of responsibility in science ultimately is a matter of struggle for power and influence regarding which ascription of responsibility should be the dominating one.

The divergence in views of responsibility is particularly obvious when it comes to the application of scientific findings. Are scientists doing basic research, e.g. gene sequencing, responsible for applications of their results made by other scientists, industrialists, politicians, etc.? Some argue that basic researchers should not be held responsible for such applications–they do not constitute a domain of responsibility for basic researchers. Others argue that applications do constitute such a domain, and stress that basic scientists should assume responsibility for the applications and that other people should hold them responsible for them. View stands against view. Different social groups and individuals have different standpoints.

Taking on Different Roles and Responsibilities

Every profession has an ethical code or practice that calls the individual to assume responsibilities beyond the ordinary. However, within a profession different individuals may assume different responsibilities. This is true also of scientists. As already pointed out, within the shared responsibility of scientists it seems reasonable that scientists high in the scientific hierarchy should take more responsibility than doctoral students. They have more experience and more power to influence. But each scientist has his or her role to play. Some scientists may assume the role of social activists trying to influence systematically the uses of scientific findings in society. Others are promoters of moral consciousness and social responsibility within the scientific community. But each individual scientist should at least assume the responsibility of being a scientist with moral integrity. However, moral integrity not only implies avoiding fraud and treating well human subjects and animals used in research, it also implies a wider social responsibility. Sooner or later there might emerge misrepresentations and misuses of his or her findings. Then, if not before, it is time for the scientist to speak out. This is even more true when it comes to the responsibility of the research teams.

The scientific community at large also has a responsibility. Moreover, within this community there are subcommunities with special responsibilities depending on their special roles. Take the human genetics community as an example. It seems reasonable to argue that this community should take greater responsibility for genetics and its applications than the scientific community at large. Moreover, when we talk about the responsibility for applications of genetics made in a medical context, it seems reasonable to distinguish between the responsibility of the community of medical geneticists and clinical geneticists on the one hand, and the community of human geneticists doing research that only very indirectly has to do with medicine on the other hand. The long term goals of such applications belong more directly to the responsibility of the former community than to that of the latter.

Political Control and Moral Responsibility

It is important to distinguish between moral responsibility and political control. Individual scientists might have a moral responsibility even if their ability to affect the direction of research is narrowly restricted by politicians and other decision-makers. A restricted autonomy is compatible with a strong individual responsibility. As Matthias Kaiser puts it in a parable (Kaiser, 1993, p. 21):

> ... each player in a soccer-team is fully responsible for his actions, in spite of having his autonomy in shaping the game being restricted by enforced rules of the game and by the capriciousness of the other members of the team.

On the other hand, this does not mean that the scientists do not have a moral responsibility to criticize certain politically inspired research priorities–of course, they do. They should not simply adapt to whatever priorities the government decides. One obvious example would be the Nazi-governed biomedical research in Germany in the 1930s and 40s. But it is equally essential in our present-day democratic societies that researchers try to influence the research priorities.

The important questions for the individual scientist are: What research should I do within the given limits? How should I do it? Should I try to change these restrictions and priorities? The individual scientist has full moral responsibility for his or her answers to these questions and for the actions performed on the basis of them.

3. RESPONSIBILITY 'TO'

We have discussed the issue of 'responsibility of', i.e. the issue of who is responsible in science. Let us now turn to the issue of 'responsibility to'. The responsibility of scientists is a responsibility to other parties involved in or affected by their research. There are many possible parties to which scientists could be responsible. Most are uncontroversial, others are disputed.

'Responding' and 'Owing'

Before we discuss who or what these parties are, let us for a moment reflect on the two fundamental metaphors that determine our thinking on responsibility, the 'response' metaphor and the 'owing metaphor (see p. 1).

The 'response' metaphor stresses that our lives are characterized by listening and answering. First comes listening, then answering. We are affected by other parties, and we react to them. In this sense, we are responsible to all entities that affect us. However, this does not mean that we are morally responsible to all of them (cf. Moran, 1996, pp. 59-65). To whom or what are we morally responsible and not merely responsible in the sense of being responsive to their influence on us? To whom or what should we respond for moral reasons? This metaphor of responsibility also has implications for our view of moral responsibility in science. To whom or what should scientists respond morally?

The second metaphor–the 'owing' metaphor–highlights another aspect of responsibility. We owe things to others and others owe things to us. What we owe each other, more precisely, is naturally an important question. But that question concerns the specific content of responsibility and will be discussed later. Here it is enough to recognize that we think of what we owe each other in terms of debts and credits, i.e. duties and rights. Other people have rights and we have duties to them. We have rights and others have duties to us. To whom do we owe things? This is also a relevant question with regard to moral responsibility in science. To whom do scientists owe things?

These aspects of 'responding' and 'owing' will be of central importance in several discussions below and in coming chapters.

Responsibility to Individuals, Communities and Organizations

It is uncontroversial to argue that scientists have moral responsibility to human subjects participating in their research and to other scientists. The main reason for this view is connected with the 'response' metaphor. Scientists interact with these parties. They influence the human subjects and these react in turn. Their response is due to their needs. It is therefore reasonable that scientists respond to these needs. Moreover, scientists influence other scientists, not only those belonging to the same research team but also scientists who take part of published results and perhaps even base their own research on these results. Sometimes the scientific community at large may also be affected by the research.

Scientists may also be responsible to different institutions and organizations, for example, to their department and university, or they may be responsible to a private company. As indicated above, this is becoming increasingly common in genetics research. Geneticists may be funded by a biotechnological company, function as consultants to a company, work in a company, or be members of the board of a company. Scientists may also be responsible to the government, at least to the extent that scientists receive education and resources from taxes. This in turn means that they may also be responsible to the general public.

It is important to note, however, that when we talk about moral responsibility to organizations including companies, what is meant is that we are morally responsible to people in these organizations, not to the organizations as such. We may talk about legal or economic responsibility to organizations but not about moral responsibility (see pp. 7-8).

Why do scientists have a moral responsibility to universities, private companies, the government, and the general public? Here the 'owing' metaphor is particularly relevant. To the extent scientists owe resources to one or the other of these parties, they have responsibility to them. This view comes close to the model of ethics in scientific research proposed by Kristin Shrader-Frechette. She calls her model 'the trusteeship model', and argues that, to the extent that scientists have society to thank for their education and research resources, they owe society research that is of value to society. Their scientific activities presuppose the social infrastructure paid for by taxes. The scientists have the trust of society to do research–they are 'trustees'–and have, therefore, a responsibility to do it for the public good (Shrader-Frechette, 1994, p. 9). I think that this model is very useful and quite in line with my general approach to moral responsibility.

However, the responsibilities to scientific institutions, companies, the government, and the general public, may come into conflict. Such conflicts of responsibility are not easily resolved. To understand the nature of such conflicts it is necessary to understand the intimate relation between values on the one hand, and communities and organizations on the other. Common values or goals constitute communities and organizations; they give a community or organization its identity. Now, an individual may belong to different communities and organizations. Take a geneticist working on the HGP, for instance. The genetics community to which he or she belongs has the common goal of understanding the genetic make-up of humans. The biotechnological company of which he or she is a consultant has the goal of making profits. The government, which provides the funding for the research from taxation of the general public, has the goal of the public good. Commonly, the values and goals of the different communities and organizations to which the scientist belongs do not conflict. Sometimes, however, they might. The scientist then faces a conflict of responsibility. It can be argued that in such situations of conflict, the scientist should give priority to the public good and become a 'whistleblower'. Being society's 'trustee', the scientist owes this to society. Choosing the public good may, however, lead to serious personal problems as the history of 'whistleblowing' clearly indicates.[6]

Controversy: Responsibility to Animals, the Environment and Future Generations?

---

[6] See, for example, Wilkinson, 1998.

The two metaphors of responsibility point in different directions. If we think of moral responsibility in science in terms of 'owing', it might seem that we only owe those who can owe us. This means that we can be morally responsible only to other moral beings, i.e. to grown-up human beings. But what about small children: are we not morally responsible to them? One possible answer is that small children belong to the moral community since they belong to the human species. A more attractive answer separates rights and duties, and maintains that a party may have rights without having any duties. This solution changes the meaning of the 'owing' metaphor. It extends the metaphor of rights. In itself, such an extension is quite possible to make. According to cognitive semantics, we make different metaphorical extensions all the time in everyday language–the question is whether it is acceptable in this particular context. However, if we accept the extension of the metaphor of rights to small children with no duties, why not extend it to animals or even the environment as well?

If we think of moral responsibility in science in terms of 'responding', the scope of responsibility can be even more extended. In a sense, we are responsible to all entities that affect us. But an adequate response is always congruent with the properties of the entity to which we respond. This seems to imply that we can be morally responsible only to some beings, not to all. What properties are relevant to moral responsibility in contrast to mere responsibility? One answer is that these entities must have at least some well-being or good of their own that can be harmed by an immoral response of humans. This suggests to many philosophers that we can be morally responsible only to sentient beings. Such beings are conscious and able to feel pain. Pain is here understood in the wide sense including stress, distress, fear, anxiety, and discomfort. On the basis of the criterion of sentience, it is obvious that scientists are morally responsible to small children, fetuses, and animals. This puts certain moral restrictions on their research. On the other hand, the criterion of sentience excludes a moral responsibility to the environment in the general sense. To the extent that that the environment or an aspect of the environment is nonsentient, scientists have no moral responsibility to it.

But there are other answers to the question of what properties are relevant to moral responsibility. Actually, the central debate in the still emerging field of 'environmental ethics' focuses on precisely this problem. Many different authors have suggested that we can talk about intrinsic value in relation to the environment and this can reasonably be interpreted as if they mean that we have a moral responsibility to the environment due to this intrinsic value. Some of them argue that all living things have intrinsic value (biocentrism), others that ecosystems do (ecocentrism). Here is not the place to present this debate. This book is mainly about the moral responsibility of scientists doing human genetics research and this kind of research does not affect the environment, other than possibly very indirectly. But, of course, in nonhuman genetics research such as that on genetically modified organisms (GMOs), the question of the possible moral responsibility to the environment may be very important. Let me just end this brief analysis by stressing the distinction between moral responsibility 'to' the environment and moral responsibility 'regarding' the environment. It is possible to deny that scientists have a moral responsibility to the environment and at the same time argue that they have a moral responsibility regarding it. In the latter case, the moral responsibility might be justified by reference to the harm destructive behavior toward the environment might bring to sentient beings, including humans. As I have argued elsewhere, sentience is quite sufficient a platform for practical environmental work (Nordgren, 1997, pp. 164-166).

Let us now turn to another controversial question, namely whether scientists can be morally responsible to future generations. This question is particularly important in gene therapy research. Would germline gene therapy be morally acceptable given the fact that it will affect future generations, not only the individual at hand? However, the question is also important to science in general, not least because it has to do with the possible moral responsibility of scientists for a 'sustainable development', i.e. a global economic development that meets "... the needs and aspirations of the present without compromising the ability to meet those of the future" (World Commission, 1987, p. 40).[7] Two important negative answers to the question whether we all, including scientists, can be morally responsible to future generations have stimulated the debate.

According to one argument, we know too little about future generations to be able to specify any responsibilities to them. We do not know who they will be, that they will be, what they will be, or what their needs and interests will be. An important objection to this argument is that we, on the contrary, have a fairly good idea about the needs and interests of future generations. This includes, for example, protection from pollution and disease.

Another negative answer to the question as to whether we all, including scientists, can be morally responsible to future generations, stresses that it is meaningless even to talk about a moral responsibility to future generations. The key point is that because different policy decisions and other decisions result in different future persons, there simply is no one future generation that would be better or worse off by either decision. The (numerically) same individuals will not come into existence since they will not originate from the same egg and sperm from which they in fact originated. Which persons will exist depends on the exact timing, and this in turn depends on policy decisions as well as other decisions (cf. Parfit, 1976, pp. 100-115).[8] One objection to this argument is that we are not morally responsible to any particular future persons but to the interests those persons, whomever they turn out to be, will have (Baier, 1984, pp. 214-215). Another objection starts from a distinction between 'possible people', those who could but do not necessarily exist,

---

[7] The 'Brundtland Commission'.

[8] See also, for example, Sagoff, 1988, pp. 60-65.

and 'future people', those who will exist in the future. It is absurd to say that we are morally responsible to merely possible people, but it is meaningful to talk about a moral responsibility to future people. The happiness and suffering of the latter can be compared. It would be irresponsible to bring into being persons who will almost certainly suffer. They need a kind of minimum quality to their lives (Warren, 1982, pp. 74-90, 154).

These objections indicate that it is acceptable in principle to talk about a responsibility 'to' future generations. However, it also seems reasonable to talk about a responsibility 'for' future generations. We are responsible 'for' them as consequences of our acts, i.e. their identity. Different decisions in the present will result in different future persons.

4. RESPONSIBILITY 'FOR': CHOICE OF SUBJECT

Let us now take a closer look at the aspect of 'responsibility for'. Scientists are responsible for the choice of subject (topic), the research process in the narrow sense (empirical and theoretical work), publication, and, more controversially, application (use) of the results. I shall start with the choice of subject. In a sense the choice of subject is the most important stage of research, because it determines in a fundamental way the direction both of research and its application. Two important general problems arise in this context.

The first general problem is to what extent the choice of subject should be under the control of the scientific community. Above I have stressed that individual scientists have a moral responsibility independently of the degree of political control. However, the problem remains as to how autonomous the scientists should be in choosing their subjects of study. To be able to solve this problem we must distinguish between different levels of choice of subject in science. The first level concerns whether there should be any scientific research at all. At a level below this we face the question about what kinds of research should be done and in what proportions. How much basic science should we have and how much applied science? Below this, how much research should be done in the natural sciences, on the one hand, and in the social sciences and the humanities, on the other? At levels further below we have the questions about what kinds of projects should be performed within the diverse disciplines and the specifics of these projects. The meaning of 'control' might diverge in these cases. At the higher levels, it is predominantly a matter of economic control, at the lower a matter of decisions of a more purely scientific character.

Which decisions should be made by the politicians and which should be made by the scientific community? Here there is a great variety of opinions. At the one extreme, there is the view that decisions in these matters should be left entirely to the scientists themselves. The scientific community should have full autonomy. At the other extreme, we have the view that the politicians should have complete control. They should decide even concerning very specific issues. Certainly, no one actually defends either of these two extreme views. Everyone adopts some intermediate position, and so do I. It is not self-evident, however, that the politicians should decide at higher levels and scientists at lower levels, as is often assumed. An alternative could be to let scientists and politicians decide together at all but the lowest levels.

Scientific or Social Value?

The second important general problem concerns the research priorities that the moral responsibility of the scientists demand, independently of their actual power to control. What should characterize a morally responsible choice?

There are at least two possible kinds of justification for the choice of subject:
- scientific value, and
- social value.

This distinction can be compared to the distinction between internal and external values of an enterprise mentioned earlier. Scientific value (or internal value) is an uncontroversial demand, although its exact meaning is a matter of dispute. The question of social value (or external value) is much more problematic. Within a specific discipline there is often a general consensus about which research projects are of great scientific value and which are not. But when it comes to social value, views are much more divergent. The well-known *Nuremberg Code* (1947), for instance, stresses explicitly the importance of yielding fruitful results for the good of society.[9] But what is 'the good of society'? The *Nuremberg Code* does not give any guidance. Looking around we find many different views representing the whole political and ethical spectrum. Take, for instance, the issue of human genetics. There are many different views about whether and to what extent human genetics is of great social value. Some think that the money could be spent in better ways. Others fear some of the social consequences of this kind of research.

Many questions arise in this context: Should scientists ever take social value into consideration in priority-setting? If so, in what way? Should they primarily do research of great expected social value or should the scientific value be the only decisive aspect? Should we have as a reasonable minimum requirement that they avoid research projects with anticipated negative social consequences? It is important to recognize that research as a pure search for knowledge can be morally acceptable. To some extent it does not need any justification in terms of social utility, at least if this utility is narrowly defined. It is a part of the process of self-understanding of the human community. A society valuing cultural richness is morally allowed to allocate at least some resources to scientific research aiming solely at increased knowledge.

---

[9] The *Nuremberg Code* can be found in, for instance, Annas, Grodin, 1992, p. 2.

However, it is also obvious that social value cannot be ignored. There are several reasons for accepting, at least partly, social value as justification for the choice of subject. The first two arguments explicitly allude to the two metaphors behind the concept of responsibility, namely the 'response' and the 'owing' metaphors (see p. 1). The first argument is the responsiveness argument. Research has a social impact, therefore scientists have an obligation to produce knowledge of social value and thereby respond to the needs of the general public. The second argument is what I call the owing argument. To the extent that scientists have society to thank for their education and research resources, as well as for the general infrastructure of society making research possible, they owe to society to do research of great social value. They have the trust of society and have, therefore, responsibility for doing research of great anticipated social value.[10] These two arguments can be supported by what could be called the historical argument and the magnitude argument. The historical argument points at the experiences from the research on the atomic bomb in the early 1940s and stresses that even if scientists during earlier centuries perhaps could have renounced responsibility for the social value of research, this is no longer possible after the atomic bomb as it has once and for all taken away the innocence of the scientists. They can no longer hide from their social responsibility. The social impact of science is much too great.[11] The potential of science is also emphasized by the magnitude argument. The technological development that has been made possible by science distinguishes itself both quantitatively and qualitatively from all earlier human activities and demands that scientists assume a responsibility for the social value of research.[12] Genetic engineering is a clear example of this. In not too distant a future, we may for the first time in history be able to genetically design our own evolution, for better or worse.[13]

An important objection to using social value as justification is that a one-sided focus on research of great expected social value might be counterproductive. History gives many examples of how socially useful results have been the outcome of research that has been performed without any such intentions. And if we allocate too much resources to research with great expected social value, we might miss some basic findings leading to even more socially useful applications. Only if the scientists are politically free to do research that they expect to be of great scientific value, independently of any expected social value, can they be creative enough to make contributions that might in the long run be of real social value.

---

[10] This kind of argument can be found in Shrader-Frechette, 1994. As mentioned above, the author calls her proposal the 'trusteeship model' (p. 9). See p. 66.

[11] This kind of argument has been proposed by Daniel Callahan in Callahan, 1976, pp. 3-4.

[12] As mentioned in the Introduction and in Chapter 1, Hans Jonas is a thinker who has stressed the novelty and potentiality of modern technology and also elaborated on the implications of this for responsibility; see Jonas, 1984.

[13] To some extent, we have always been able to design our future, for example by mate selection. But genetic engineering adds a qualitatively new dimension.

I agree that a too one-sided focus on research with great expected social value might in the end be counterproductive. But this does not imply that we should not focus a great deal on such research. If we see social or ecological problems that have to be solved and we have good reasons to believe that science might make an important contribution, we should naturally focus on such research. One example could be research on global survival and sustainable development (Nordgren, 1997, pp. 162-163).

Another important objection stresses the difference between basic and applied science. These two kinds of science have different objectives and, consequently, different implications for the question of responsibility. The aim of basic science is to understand, while the aim of applied science is to influence. The implications are that while scientists doing applied research are responsible for doing research of great potential social value, scientists doing basic science are not. Basic scientists should only use the criterion of scientific value.

This argument from a separation of basic and applied science also includes an important observation. There are clear examples of basic and applied science, and to the extent the distinction can be sustained it is of some relevance for the question of responsibility. However, as stressed above, very often the border is not at all clear and this is also something that is relevant to the question of responsibility. First, the time lag between basic findings and their application is often very short. A clear example is genetics research. Soon after a disease gene is discovered, a genetic test for clinical use becomes available.[14] Second, possible applications are often cited as justification when applying for grants to basic research. Genetics is a good example here as well. In such cases, basic science might be said to entail applied science. Third, applied research is often not only an attempt to affect humans, animals or plants. It might at the same time give us increased basic understanding. It may also define new questions to be addressed by basic research. In other words, applied science often entails basic science. Fourth, basic science often uses tools developed by applied science. All this indicates that the distinction between basic and applied science often is not clear-cut. This fact in itself has relevance for the question of responsibility. In many kinds of research, it is no longer possible to hide behind the distinction between basic and applied science. Scientists doing such basic research cannot escape responsibility for the social value of their research by citing the separation of basic and applied research.

My conclusion is that scientific value and social value are both relevant when it comes to morally responsible priority-setting in research. I therefore suggest a balancing approach in line with the imaginative casuistry proposed in the previous chapter. However, the balancing of research between projects of great potential social value and projects of only great potential scientific value is a delicate thing. Here it is obvious that responsibility is a matter of judgment from case to case, not

[14] Gene-based therapies, on the other hand, may often be late, giving rise to what is called the 'therapeutic gap'.

of following strict rules put forward once and for all. Perhaps we can see the priorities swinging as a pendulum and perhaps this swing is morally acceptable. When research has been in general too focused on social utility for some time, we might need to let it be more focused on scientific value. Conversely, when it has been in general too focused on scientific value, it might be necessary to make it more directly socially relevant. Such a swinging can be quite acceptable.

However, a reasonable minimum demand should be to avoid research projects with anticipated negative social consequences. The criterion of social value is not only a matter of high or low value; it is also a matter of positive and negative value. If a project is predicted to have a negative social value, i.e., to have negative social consequences, and no positive social value, it should not be chosen, even if it is expected to be of great scientific value. This view is clearly stated in the *Uppsala Code of Ethics* from 1984. This code focuses particularly on the responsibility of individual scientists to continuously review their own research and abstain from research with anticipated bad consequences. It stresses, for instance, that research should be given such a direction that its applications will not cause essential ecological damage or be used for war or oppression (Gustafsson et al., 1984, p. 312). To be sure, this view can also be applied to the scientific community at large. A difficult but very common problem, however, is when the possible social consequences of a research project are both positive and negative. Such projects should only be chosen if the negative consequences can be anticipated to be avoided with a high degree of certainty. Here arise, of course, difficult balancing problems.

5. RESPONSIBILITY 'FOR': RESEARCH AND PUBLICATION

The next two stages of research is the research process in the narrow sense (empirical and theoretical work) and the publication of scientific findings. Moral responsibility for these stages exhibits different aspects. I find it fruitful to make the following distinctions, which concern the scientists' actions and their direct consequences:

- moral responsibility for avoiding scientific misconduct,
- moral responsibility for maintaining high scientific quality, and
- moral responsibility for actions affecting human subjects and animals used in research, and for actions affecting the environment to the extent it is affected by the research process in the narrow sense.

Scientific Quality

The first two aspects of moral responsibility are intimately linked to the purely scientific responsibility. The scientific quality of research may have an impact on its ethical quality. Low scientific quality, including scientific misconduct, bias, bad theoretical and empirical work, may harm people, and a moral reason for maintaining high scientific quality may be a concern for these parties. With 'scientific misconduct' I mean fabrication, falsification, and plagiarism in research and publication. 'Biased' research is research shaped by ideological concerns, for example related to class, gender or 'race'. Bias might affect the questions asked, the data produced, or the assumptions on the basis of which the research is carried out. Low quality might also be a matter of 'scientism', i.e., of overinterpretation of scientific findings.

It is obvious that low scientific quality may harm scientists as well as society at large. Fabricated data harms the work of other scientists who base their own studies on these 'results'. Plagiarism violates the rights of other scientists who do not get fair recognition of their work. Scientific results of low quality also give politicians and other decision-makers a bad point of departure for their decisions.

Moral responsibility for scientific quality is particularly important in research on issues that are heavily politically charged. Genetics and gene technology are such issues. In this kind of research the maintenance of high scientific quality implies avoiding ideological bias and overinterpretation. There might emerge a temptation to manipulate scientific findings in order to promote a certain policy. This is, of course, unacceptable. In order to avoid this, it is important to distinguish between commonly accepted 'facts' and ideological overinterpretations, at least as far as possible. The problem is, of course, that there are no uninterpreted facts. What we call 'facts' are always those within a certain conceptual framework. However, this does not mean that scientists should not try to avoid ideological overinterpretations of their results. At the very least, they should explicitly state when they are expressing controversial ideological positions and leaving the area of broad consensus. An example of overinterpretation in relation to human genetics are the ideas presented in the controversial book *The Bell Curve* by Richard Herrnstein and Charles Murray (Herrnstein, Murray, 1994).[15] Other examples of overinterpretation and misconstruction of science are discussed by Alan Sokal and Jean Bricmont in their book *Intellectual Impostures* (Sokal, Bricmont, 1998). This activity of dressing politics and ideology in scientific clothes should be strongly counteracted by the scientific community. But this does not, on the other hand, mean that science has no relevance for politics and ideology. It certainly has.

Here a comment on the 'social constructivism' might be in place. A key question raised by social constructivism and its criticism in books like *Higher Superstition* by Paul L. Gross and Norman Levitt is the extent to which scientific knowledge is objective (Gross, Levitt, 1994). The views of objectivity in science can be illustrated

---

[15] For an analysis, see pp. 111-114, 227, 234.

by a spectrum. At one extreme, we find the view of radical social constructivists. In their opinion, science gives no objective knowledge whatsoever. A scientific narrative is just one narrative among others giving no knowledge of the world as it is. Perhaps there is no real world at all. Everything is 'text'. At the other extreme, we find the view that science gives a literally true picture of nature. Most scientists and scholars are probably found somewhere in between these extremes. Without a lengthy argument, let me just make it clear that my own standpoint is closer to the objectivist extreme than to the social constructivist. Note, that this is completely compatible with the 'contextual interactionism' proposed above. A recognition of science as a social activity by no means implies a social constructivism regarding scientific knowledge, but it does commit oneself to a rejection of a naive objectivism.

The responsibility for maintaining scientific quality is clearly described in the *Nuremberg Code*, mentioned above, and also in the *Helsinki Declaration* (1964, 1975, 2000).[16] However, one problem is: exactly what standards should be accepted? The codes do not give sufficient guidance. Everyone agrees that misconduct and bias should be avoided, but there might be some disagreement about where exactly to draw the line. Some theorists of science argue that what counts as 'good science' might to some extent be field-dependent and even change over time.[17] Moreover, many would argue that what is important is not ethical codes and their regulations of scientific behavior but the traits of character of the scientists. From this perspective, moral responsibility is primarily a matter of cultivating personal virtues like honesty and integrity.

Human Subjects, Animals, and the Environment

The treatment of human subjects is another important part of responsibility in science. Particularly in biomedical research the importance of obtaining the voluntary, informed consent of the research subjects has been central for a long time, as can be seen in the *Nuremberg Code* and the *Helsinki Declaration.* The *Nuremberg Code* was a reaction against the experiments on human subjects in the German concentration camps, and the *Helsinki Declaration* represents a further development and modification of its views. Neither code, however, gives guidance enough to the more specific problems that emerge in different kinds of research, although the *Helsinki Declaration* makes some effort in that direction by distinguishing between therapeutic and nontherapeutic research. Not least in human genetics research, which is the main topic of this book, special problems arise with respect to informed consent.

---

[16] See, for instance, Annas, Grodin, 1992, pp. 331-333.
[17] See, for instance, Segerstrahle, 1994, pp. 100-104.

The main argument for obtaining the voluntary, informed consent of human participants in research is respect for autonomy. Autonomy is without substance if the human subjects do not receive information enough to be able to reach well-considered decisions whether to participate. Another reason is the principle of nonmaleficence. Without informed consent human subjects might be harmed (although they might, of course, be harmed in spite of it, too!).

Alongside treating human subjects well, there is also a growing awareness of the responsibility for the treatment of animals used in research and for the impact of scientific experiments on the environment. This is clearly illustrated by the revision of the *Helsinki Declaration* made in 1975. No such references are found in the original version of 1964, but in the revised version of 1975 and in later revisions, we find a passage referring to animals and the environment.[18] The ethical committees for animals used in research are other recent manifestations of this growing awareness.

Thus, there is wide agreement that scientists have a moral responsibility for the treatment of laboratory animals and for the impact on the environment. However, one problem is what this responsibility means more exactly. What are we allowed to do to the animals and the environment and what are we not allowed to do? The *Helsinki Declaration* does not give a clear answer. Scientists assuming responsibility for the animals and the environment have to make these specific judgments themselves. Moreover, there are many different opinions on what arguments should be offered as justification for these responsibilities. In Chapter 5 I will discuss both the issue of treatment and the issue of justification with respect to animals. A discussion of environmental responsibilities, however, is beyond the scope of this study.

6. RESPONSIBILITY 'FOR': APPLICATION

I have discussed the responsibility for the direct consequences of the research process. I shall now turn to the responsibility for the indirect consequences of research that occur when the findings of research are put to use in various contexts. It is important to distinguish this aspect from the responsibility for the choice of subject. Scientific findings might be put to uses that were not intended or even anticipated when the research project was first chosen.

Like the aspects of responsibility discussed in the previous sections, the responsibility for the uses of scientific findings is stated in some ethical codes. Of special interest is the *Uppsala Code of Ethics* already mentioned in the section discussing the responsibility for the choice of subject. Here the responsibility for interrupting one's own research if one anticipates bad consequences is explicitly stressed. This is, of course, the most direct way of taking responsibility for the

[18] See, for instance, Annas, Grodin, 1992, pp. 331-342. See also URL: www.wma.net.

consequences and applications of research. In addition, the *Uppsala Code* also includes a passage that stresses the value of promoting good applications.

The question of responsibility for applications of research is particularly important when it comes to human genetics. A not unimportant ethical declaration regarding human genetics, the *Declaration of Inuyama*, states clearly that geneticists have a responsibility for the applications (CIOMS, 1995):

> In terms of ethics and human values, what must be assured are that the manner in which gene mapping efforts are implemented adheres to ethical standards of research and that the knowledge gained will be used appropriately, particularly in genetic screening and gene therapy... Genetic researchers and gene therapists have a strong responsibility to ensure that the techniques they develop are used ethically.

But what do we mean by 'applications'? The answer depends to some extent on whether we talk about applications of results in basic science or applications of results of applied science. I find it useful to make the following distinctions:
- uses in policy contexts and other social contexts of the results of basic science,
- uses made in applied science, and
- uses in industrial, clinical, policy, and other social contexts of applications developed in applied science.

It is also important to distinguish applications made by oneself and applications made by other people, including other scientists, politicians, industrialists, etc. This distinction is directly relevant to the issue of moral responsibility for applications. Taking this distinction as a starting point we may talk about:
(1) being responsible for one's own applications,
(2) being responsible for other people's applications of one's own results, and
(3) being responsible for trying to affect other people's applications of one's own results.

The first alternative is unproblematic and will not be discussed any further. Actually, it is part of the responsibility for the research process in the narrow sense and publication, both of which have been discussed in the previous section. Alternatives (2) and (3), however, raise several problems which I intend to discuss rather extensively.

But before I turn to this discussion let me briefly comment on the theoretical status of science-based technology. Basically, there are two different models (cf. Bereano, 1997, pp. 132-133). The first can be called the 'use/abuse' model. Here, technology is understood as neutral in itself but something that can be used or abused. According to the second model, technology always reflects underlying relationships of power. Powerful sectors in society are able to articulate research programs, get funding for them, carry them out, and develop technologies on the basis of the scientific findings. The second model seems more adequate. As we shall see in Chapter 3, a very clear example is the Human Genome Project, not least in the US. This is well illustrated already in the title of the book on the origins of the HGP by Robert Cook-Deegan, *The Gene Wars* (Cook-Deegan, 1994). This view on technology has implications for bioethical reflection. In the assessment of gene technology, it becomes meaningless to discuss the ethical acceptability of this technology in abstract and 'in itself'. Instead it becomes necessary to discuss it with an eye on the context, i.e. the social power relations. Note that this is quite in line with the view on the relation between science, technology and society, proposed in the beginning of this chapter, namely contextual interactionism, as well as with the proposed modified social model of moral responsibility.

Responsibility for other people's applications?

Alternative (2) raises the problem of whether one can be responsible for other people's actions. It seems possible to assume responsibility for other people's actions, although this is quite unusual except in the case of parents assuming responsibility for the actions of small children. If one assumes responsibility for other people's actions, then one can also reasonably be held responsible for them by other people. This can be the case, even if one goes far beyond one's ordinary responsibility. If we apply this to responsibility in science, it appears that it is possible for scientists to assume responsibility for other people's applications of their findings. If so, other people can reasonably also hold the scientists responsible for them. This assumption and ascription of responsibility for other people's applications can be both backward-looking and forward-looking. The applications can be uses already made in the past or uses that will be made in the future.

However, there is also another angle to this. Other people's applications can be viewed as consequences of scientists' publication of their findings. It can reasonably be argued that we are responsible for the consequences of our actions, at least to some extent. But to what extent? We can think of this in terms of the metaphor of a spectrum of responsibility ranging from the extreme of full responsibility to the other extreme of no responsibility at all.[19] Let us take the responsibility of scientists doing basic research as an example, although that of scientists doing applied research can be imagined in quite similar a manner. The suggestion should be seen as a normative proposal.

Scientists have full and direct responsibility only for their own acts and acts of omission. However, they may also have responsibility for applications of their results, made by other people, viewed as consequences of publication. It seems reasonable to argue that the degree of responsibility depends on intention, predictability, and conceivability. If the applications made by other people are foreseen and intended by the scientists doing basic research, they have indirect but nevertheless almost full responsibility for them. If the applications are foreseen but not intended, they have great but not full responsibility. If they are unintended and unpredicted but conceivable in principle, the responsibility is more limited, but

[19] Note that this is a metaphor. It is not possible to make a real quantification of responsibility.

perhaps not zero. Perhaps the scientists could have and should have tried to get the information needed to be able to envision them? If no applications can be conceived at all the basic researchers have clearly only a very indirect causal responsibility for them if they nevertheless would come. It can be argued that they have no or almost no moral responsibility for them. They should neither be blamed nor praised for these applications.

Responsibility for Trying to Affect Other People's Applications?

Scientists are only indirectly responsible for applications of their findings made by other scientists, industrialists, politicians, etc. They have only indirectly causally contributed to these applications by publishing their findings and this is a good reason for arguing that they, therefore, are only indirectly morally responsible for them and this to a varying degree. However, what is truly important when talking about the moral responsibility for other people's applications is not whether scientists should be blamed or praised for other people's applications. The important thing is that scientists actively take responsibility for trying to affect other people's applications. Therefore, for the remainder of this section I shall focus rather extensively on this third alternative.

According to a traditional view, scientists have no responsibility for trying to affect the applications of their findings. The question is whether this traditional view is acceptable today. Perhaps the responsibility for the choice of subject, research and publication should be supplemented to include even application? Perhaps the responsibility of the scientists should be expanded? From the perspective of the proposed modified social theory of moral responsibility, this is at least a possibility. Scientists can assume a responsibility that is not traditionally theirs. This is possible because moral responsibility is a matter of ascription and assumption by individuals and social groups, not simply a matter of causal agency. But such an expansion of the responsibility of scientists is rather controversial. Many different arguments can be put forward for and against this view. Let us begin with an analysis and discussion of some of the main counter-arguments. Note that in order to facilitate ethical deliberation I proceed rather systematically.

Arguments Against

I find at least five important arguments in this respect. The argument from noncontrol stresses the insufficient ability of the scientists to direct the uses of their results made by other people. If they cannot control the uses, they cannot reasonably have a responsibility for them.[20] The argument from value-freedom points out that science is concerned with facts, not with values. And since the use of the results of

[20] This argument is referred to in Callahan, 1976, p. 1.

scientific research is a matter of values, it falls outside the domain of competence of the scientists.[21] The argument from lack of consensus stresses that even if scientists should assume responsibility for trying to affect the applications of their results made by other people, they may not agree on which applications are good and which are bad. There may be a plurality of different views. Therefore, they should abstain from trying to affect the uses in a certain direction. The role argument starts from the assumption that different professions have different tasks to fulfill in society. If and only if the professions stick to their roles and to the tasks that go with these, society functions well. If the scientists doing basic research try to affect the use of their results, they confuse their role with the roles of politicians, industrialists and clinicians, and this is something that society at large loses on.[22] The argument from limited time is pragmatic. It acknowledges that it would be a good thing if scientists tried to affect the applications of their findings but stresses that such a concern takes a lot of time and effort. There will be too little time for the scientists to do what is their primary responsibility, namely to do research. The argument from a separation of basic and applied research, which we in fact have met in the discussion of the choice of subject, stresses that these two kinds of research have different objectives and, consequently, different implications for the question of responsibility. The aim of, for example, basic genetics is to map and sequence the genome, and understand the functions of different genes, while the aim of applied genetics is to apply this knowledge in clinical and other contexts. Scientists doing basic genetic research are not responsible for applications of their findings made by other people. Scientists doing applied genetic research are responsible for their own applications. None, however, are responsible for the uses in society of these applications.

How should these arguments be assessed? Let us start with the argument from noncontrol. To be sure, it is to some extent valid. It underlines the importance of stressing the distinction made earlier between 'being responsible for other people's applications' and 'being responsible for trying to affect other people's applications'. Scientists cannot be responsible for consequences in terms of applications of their results made by other people, other than indirectly. By publishing their findings they have contributed to a chain of events leading to certain practices in society. Their research is a necessary, causal condition for these applications. However, the view discussed is that scientists have a responsibility to try to affect applications made by other people. Admittedly, success might be very difficult–but their responsibility is to try, not necessarily to succeed. Even if the scientists cannot be held responsible, other than indirectly, for the applications of their results made by other people, they can assume responsibility for trying to affect such applications. They can try to affect future uses of their findings. They can also interrupt research that is anticipated to have negative social consequences.

---

[21] For an eminent historical presentation of the notion of a value-free science, see Proctor, 1991.

[22] The concept of role responsibility is discussed in Hart, 1968, and May, 1992, pp. 163-183.

The argument from value-freedom is also to some extent acceptable. Science is ideally aiming at establishing facts, not values. But this does not mean that science as it is actually practiced is value-free. Nor does it mean that scientists should not assume moral responsibility. It is quite possible to argue that science is aiming at establishing facts and at the same time argue that scientists should assume responsibility for trying to affect the uses of their findings in society.

The argument from lack of consensus stresses something important that should not be ignored. The scientific community may not reach full consensus regarding which applications to promote and which to counteract. However, it may reach some consensus and that is all that could be hoped for. In addition, it can be argued that each individual scientist has responsibility to act in accordance with his or her own ethical values.

The role argument also carries some conviction. There is a danger that if scientists publicly defend a certain moral point of view, they actually export some of their purely scientific authority to the field of ethics. However, this danger can be avoided by encouraging open discussion of the tenability of their arguments. Moreover, there are differences between different professions. It is wrong to compare, for instance, scientists with attorneys. The special role of attorneys has to do with the conflict between their clients and the legal system of society. Attorneys should defend their clients, not the interests of society at large. The special role of scientists does not concern such conflicts. Scientists doing basic research aims at understanding some aspect of nature. Certainly, as mentioned above, there might emerge conflicts, for instance, between the interests of scientists working for private companies and the interests of society at large. But such conflicts have nothing to do with the scientific role itself. It is quite possible to argue that scientific research should be carried out for the public good. This means that scientists should assume responsibility for trying to affect the use of their findings, not hide themselves behind a narrowly defined social role.

The argument from limited time is also partly convincing. Efforts to influence the direction of applications might take a great deal of time and the primary responsibility of the scientists is to do research. However, responsibility for trying to affect the applications is not necessarily the responsibility of each and every individual scientist, at least not a full time responsibility. But some scientists might assume such a responsibility for a limited period of time. However, the overall responsibility lies with the scientific community at large. It should see to it that some scientists get the opportunity to spend time trying to affect the applications.

The argument from a separation of basic and applied science also makes an acceptable point, at least when stated in general terms. There are clear examples of basic and applied science, but very often the border is not at all clear. As we have seen above, the time lag between basic finding and application is often very short, possible applications are often referred to in grant applications, and applied research may define new questions to be addressed by basic research. This indicates that the distinction between basic and applied science is fuzzy. This is relevant to the issue of moral responsibility. Basic science researchers cannot escape from the responsibility of trying to affect the uses of their results by referring to the distinction between basic and applied research.

Arguments in Favor: Responding and Owing

The *con* arguments, against the view that scientists doing basic research should take responsibility for trying to affect the applications of their results, are not entirely convincing. What arguments can be put forward in favor? Actually, the *pro* arguments are almost the same as those put forward for the view that scientists have a responsibility for choosing study subjects of great social value.

According to the responsiveness argument, scientists have a responsibility to respond to the needs and fears of the general public, because their research has social impact. This means that they should try to promote positive uses of their findings and counteract negative uses. They should do this not least in order to show that they take the fears for the consequences of a particular technological application seriously. According to the owing argument, scientists owe to society to assume responsibility for trying to affect the applications of their findings, because they have received from society such things as education, research resources, as well as the general infrastructure that makes research possible. They have the trust of society and should, therefore, assume responsibility for trying to affect the uses of their research.[23] The historical argument points at the experiences from the research on the atomic bomb in the 1940s and stresses that even if scientists during earlier centuries perhaps could have renounced responsibility for the consequences of research, this is no longer possible after the atomic bomb. Science is much too powerful and may influence society radically. The potential of science is also emphasized by the magnitude argument. The technological development that has been made possible by science distinguishes itself both quantitatively and qualitatively from all earlier human activities and demands an expanded responsibility. The emergent technology of gene therapy is a clear example of this. If research on this kind of technology becomes truly successful sometime in the future, human beings will become able to change their own genetic make-up in a fundamental way never seen before, for better or worse.

The historical argument must be taken seriously. The experiences of the atomic bomb have once and for all removed the innocence of the scientists. In genetics, the historical experiences of eugenics during the first decades of the 20th century clearly indicate the need for moral responsibility. The magnitude argument is also very compelling. In particular, genetic research seems to have a potential never seen before in the history of mankind. The scientists cannot run away from

---

[23] Cf. Shrader-Frechette, 1994, p. 9. See pp. 66, 71.

their responsibility. However, the decisive arguments are the responsiveness argument and the owing argument. Scientists should respond to the needs and fears of the general public. Moreover, they owe to society a responsibility for trying to affect the application of their research. This is also true of scientists working in private companies. Even if their more specific resources come from their company and not from the state, they are dependent on the general infrastructure of society.

The conclusion of this evaluation is that the *con* arguments are not acceptable, while the *pro* arguments are. However, we have found that there are some valid points in the counter-arguments. In particular, the argument from noncontrol indicates that it is reasonable to emphasize the distinction between 'being responsible for other people's applications' and 'being responsible for trying to affect other people's applications'. Even if the scientists cannot reasonably be held responsible for applications of their findings made by other people, other than indirectly, they should actively take responsibility for trying to affect them.

So far, I have presented four general arguments for the view that scientists should assume responsibility for trying to affect applications of their results. However, this book is mainly about human genetics; are there any special reasons for geneticists to assume such responsibility? Actually, we have already met two such reasons presented as objections to the argument from a separation between basic and applied science.

According to the first argument, the time lag between basic finding and application is often short in genetics. Geneticists doing basic genome research can often anticipate possible applications. The second argument stresses that possible applications are often referred to as justification when applying for grants to basic genetic research. Often possible medical applications are referred to. Both these reasons give special support to the view that geneticists should assume responsibility for trying to affect applications of human genetics made by other people, including other geneticists, industrialists, politicians, etc. Not least does it become important for geneticists to reflect on the long term goals of the application of genetics in society.

In conclusion, there seem to be good reasons for scientists in general and geneticists in particular to assume responsibility for trying to affect the applications of scientific results. However, should each and every scientist assume such responsibility? I do not think so, at least not as a full time responsibility. I have already stressed in my comment to the argument from limited time that the overall responsibility lies with the scientific community as a whole. The scientific community should make it possible for some scientists to spend time trying to affect the applications. These scientists should take upon them the special role of promoting good applications and warning against bad applications. Concretely, this could imply starting a discussion within the scientific community about these issues as well as opening a dialogue with the general public and the politicians.

7. CONTENT AND FORM OF RESPONSIBILITY: GENERAL PROPOSALS

I have discussed some arguments concerning different domains of moral responsibility in science. A domain of responsibility is a crossing of a stage of research (*for* which scientists are responsible) and a party (*to* whom scientists are responsible). I have argued that scientists should take responsibility *for* the choice of subject, the research process in the narrow sense, and the publication, as well as for trying to affect the application of their findings. I have also argued that scientists are responsible *to* other scientists, human subjects, animals used in research, the public, the government, and, in some cases, private companies. In addition, I have argued that scientists are also responsible *to* future generations whoever they turn out to be. With respect to the environment I have been more hesitating. Without giving a lengthy argument–the issue is not of central importance to this book, which is primarily about human genetics–I have indicated that it seems more adequate to say that scientists are responsible 'regarding' the environment than 'to' it.

Now, I shall turn to the content and form of moral responsibility in science. My proposals will be rather general. In coming chapters, where I will focus exclusively on human genetics, I shall be more specific. My normative suggestions are based on the view presented in the previous chapter, namely 'imaginative casuistry'. The targets of my proposals are those scientists who are prepared to take responsibility not only for their choice of subject, the research process in the narrow sense, and publication, but also for trying to affect other people's applications of their findings. In the remainder of this book, I call these scientists 'responsible scientists' or–more specifically, when I talk about human genetics–'responsible geneticists'. The proposals should be understood as a contribution to a dialogue.

What should responsible scientists do? My general proposals to these scientists–as individuals and as a community–are as follows:

(1) Use your moral imagination to envision different ethically relevant consequences of research, and to figure out different ways of taking responsibility for these consequences.

(2) Learn from history, i.e. from earlier, prototypical cases.

(3) Participate in a dialogue with the general public, politicians and industrialists.

(4) Integrate ethical reflection with scientific practice by choosing an appropriate form of responsibility, i.e. adequate means of implementing the content of responsibility at different social levels.

Let us take a closer look at these proposals.

Moral Imagination

For responsible scientists, the use of moral imagination is fundamental. First, it should be used to envision different ethically relevant consequences of their research. How are different parties affected by different stages of research including application? To some extent this question can be answered by empirical investigation. In genetics, the ELSI research is an attempt to do this. However, with regard to long term consequences, this may be difficult. In this case, the scientists have to use their scientific imagination to envision possible future developments and consequences. But this is not sufficient. Scientists also need to determine which possible consequences are ethically relevant. They need to empathize with those affected by research and its application. This requires a moral sensitivity.

Moreover, moral imagination should be used to figure out different ways of taking responsibility for the consequences. Scientists need to find creative ways of assessing the anticipated consequences. Which consequences are ethically acceptable and which are not? This implies an analysis of different arguments in favor and against the different options. These arguments often refer to ethical principles such as autonomy, utility, justice, and integrity. Their metaphorical character need to be made explicit. Different interpretations of these principles need to be noticed and prototype cases that support the different interpretations need to be identified. Their scope of application has to be recognized. How far are the metaphors extended to nonprototypical cases? Finally, the relative weight of different principles should be detected.

In addition to carrying out this analysis, responsible scientists should take a stand. Which consequences should be promoted and which should be counteracted? Which interpretations, applications and rankings of the principles should be made? When they have taken a stand, they should try to find creative ways of implementing their views. This means that they should try to integrate their ethical views with their research and social activity. Which forms of responsibility are the most appropriate? With regard to applications of their findings, how should they attempt to influence the use of their findings in society?

To be sure, moral imagination is vital at the level of personal research ethics. The principles of the ethical codes of scientific conduct are often too general to give sufficient guidance. There is always a need for making judgments in particular cases and this should be explicitly recognized when we talk about the moral responsibility of scientists. In all such judgments, moral imagination plays a key role.

However, moral imagination is also essential at the level of policy and law. Scientists should assume responsibility for policy-making and take into account aspects of broad social concern. This is not excluded by a casuistic approach, something that is clearly shown by Baruch Brody in his book *The Ethics of Biomedical Research: An International Perspective*.

In this book, Brody analyzes research policy documents from different organizations and countries around the world and finds a clear avoidance of fundamental theories employing single values. Instead, many of these documents show a balancing of a variety of legitimate but conflicting principles and values (Brody, 1998, pp. 202-206).

Brody also finds that the official policies of research ethics all emerged according to a common pattern. They arose in response to (1) controversial cases, (2) new technology, or (3) broader social concerns with implications for research. Controversial cases such as the Nazi experiments and the Tuskegee Syphilis Study led to policy statements about the treatment of human subjects in research. New technology such as genetic engineering and gene therapy has led to official statements concerning the application of scientific results. Broader social concerns have also influenced research ethics. A broader concern about animal rights has had implications for animal research and a broader concern about individual rights and mistrust of authorities has had implications for individual consent in research with human subjects, just to mention two examples (Brody, 1998, pp. 197-199).

Finally, Brody finds a high degree of convergence in the policy statements. On some issues there has been a remarkable consensus–one example is the treatment of human subjects in research. On other issues there is a still emerging consensus; he mentions genetics research and its application as an example. However, there are also some real disagreements, for example, regarding animal research (Brody, 1998, 199-202).

Brody's conclusions seem well founded. The official policies are best understood as attempts to balance commonly accepted but conflicting principles and values in response to new circumstances. It is obvious that the policies are case-driven, not theory-driven. They are articulated in response to controversial cases, new kinds of technology or broader social concerns, not as context-independent expressions of fundamental single-valued theories. Moreover, this is quite in line with the thesis in the previous chapter that some kind of casuistry is the most psychologically realistic normative position. Not only in everyday life but also in policy-making regarding scientific research a casuistic methodology is to be expected. In addition, it is worth noting that Brody's analyses support the conclusion by Jonsen and Toulmin from the work on US policies on human subjects research in the 1970s which I also mentioned in the previous chapter. Brody has found a much more general pattern.

Lessons from Historical Cases

One very important point in taking moral responsibility for scientific research is to learn from history, i.e. from earlier, prototypical cases. According to imaginative casuistry, prototypical cases are at the very core of moral reasoning. This holds for reasoning in research ethics as well; research ethics is case-driven.

Historical awareness is important at all stages of research. However, it plays a key role when it comes to the application of scientific findings. I have already referred to the atomic bomb and its implications for demolishing scientific innocence in general. Historical cases can be referred to in arguments for assuming responsibility for a certain domain, in this case for the application of scientific findings. But history can also be used in arguments regarding the content of responsibility. For instance, when we are talking about the content of the responsibility for the applications of genetics other historical cases than the atomic bomb are more directly relevant. First, I am thinking, of course, of the discipline's eugenic past (cf. Beckwith, 1993). Here Nazi eugenics stands out as a warning example. However, more interesting cases are the versions of eugenics implemented in the US (Kevles, 1995 (1985)), the UK (Thom, Jennings, 1996), and the Nordic countries, in the first decades of the 20th century (Broberg, Roll-Hansen, 1996). I will return to these examples later. Second, much can also be learned from the 'recombinant-DNA debate' in the 1970s. In his analysis of this debate, the science historian Charles Weiner finds that technical issues were separated from the ethical, and that the ethical questions were defined as a concern for society, not for science. Weiner stresses that such 'compartmentalization' must be avoided (Weiner, 1994). He is right. The key words should be 'integration' and 'dialogue'. Let us first discuss the issue of dialogue.

Dialogue

Dialogue is an essential part of moral responsibility in science. It is a way of responding to the needs of other members of society that are affected by research, and 'responding' is a key metaphor of the concept of responsibility. This responsibility to communicate is often understood as a responsibility to inform. However, the giving of information must never be a monologue. It has to be part of a dialogue. The scientific community has a moral responsibility for opening a dialogue with people involved in or affected by research.

The most commonly stated aspect of this 'communicative responsibility' is to provide information to human subjects participating in research. There is general consensus regarding the importance of obtaining the informed consent of such participants. What has not always been stressed in this context is the need for dialogue, not just the giving of information.

However, the scientific community also has a broader communicative responsibility. It has a responsibility to participate in a dialogue with the public and the decision-makers about all the issues discussed in previous sections: choices of subject and priority-setting in research; the research process in the narrow sense and its impact on human subjects, animals and the environment; published findings; and potential and actual applications of scientific findings. It is vital that the scientists not only inform about and discuss scientific aspects, but also, when required, inform about and discuss ethical aspects.

The communicative responsibility is primarily a responsibility of the scientific community. Individual scientists seldom have the power to control the direction of research and the uses of research. However, the scientific community has definitely a responsibility for opening a dialogue with the general public and the decision-makers. It is particularly important that it takes seriously public opinion on different kinds of scientific research and on different uses of scientific findings.

Four things should be kept in mind in this dialogue. First, scientists often talk about the importance of educating the public, giving people 'scientific literacy', but, as Weiner emphasizes, it is equally necessary that the scientists obtain 'public literacy', in the sense that they understand the public and its attitudes towards science and its application (Weiner, 1994). An example of the importance of scientists obtaining 'public literacy' has to do with the reasons behind the resistance among large parts of the public against certain applications of gene technology, not least the genetic modification of animals. Surveys have shown that the main reason behind this resistance has to do with the violation of the natural order and the integrity of Nature rather than with risks.[24] Genetic modification of animals is viewed as morally unacceptable in itself, not only as something dangerous. Thus, when geneticists speak in terms of risks, it appears that they to some extent do not address the real concern of large parts of the public. A real dialogue between geneticists, other scientists and scholars, politicians, representatives of biotechnological companies, and the public, is needed, and a condition is that the scientists make a real effort to try to understand the points of view of these other parties.

Moreover, it is important to be aware of the power-relations affecting the dialogue. Different groups, for instance, universities, companies, political parties and people in the third world, may have diverging interests. Struggle for power might obstruct the possibilities for dialogue. A dialogue that transcends groups and contexts may nevertheless be possible. Scientists must at least try to open such a dialogue.

It is also vital not to be in a hurry when communicating. A view that is not the product of deep reflection is not desirable. The participants in the dialogue must take time to test many different arguments in a thorough manner. In this way, the dialogue between the scientific community, the general public and the decision-makers might be educating and transforming for all those involved.

Finally, ideally the goal of the dialogue is consensus within the scientific community and within society as a whole. However, in the real world this will probably never be achieved. How should such a situation be handled? This depends on what issues are discussed and whether we are talking about the scientific community or the society at large. Later I will argue that on societal issues which are heavily existentially loaded and regarding which there is a clear lack of consensus, e.g. issues in reprogenetic medicine, individual autonomy is extremely important. This does not mean, however, that there should be no governmental restrictions. But it does mean that these restrictions should be limited in scope. In other societal cases, democratic majority decisions in the parliament may be preferable.

---

[24] See, for example, Biotechnology and the European Public Concerted Action Group, 1997.

Regarding disagreement within the scientific community, there are two options depending on what issue we are talking about. The first is to leave the decisions to the individual scientists and not lift the issue up to the communal level. The other alternative is to let the government decide on the matter.

Integration of Ethical Reflection with Scientific Practice: Forms of Responsibility

Ethical reflection of scientists in dialogue with the general public, politicians and so forth, must be integrated with scientific practice. Take, for example, the creation of ELSI programs linked to the Human Genome Project. These programs are, of course, good. They are ways of taking responsibility in science. But, as Weiner stresses, if the results of the ELSI programs are not integrated with the research itself by influencing policies and priorities, then technical and ethical issues become separated once again just as in the recombinant-DNA debate in the 1970s (Weiner, 1994). The 'compartmentalization' must be avoided. In particular, the long term goals of scientific enterprises must be discussed and integrated.

How should ethics be integrated with scientific practice more precisely? This is the issue of the forms of responsibility. It is important to choose the adequate means of implementing the normative content of responsibility in science at different social levels. I find it useful to distinguish among four levels. The first concerns individual scientists. The ideal is that each one cultivates the virtue of responsibility in their work and life. However, there also lies a responsibility with the research teams. The members of the research teams have a shared responsibility for the 'moral climate' in the team. At the third level we find a similar shared responsibility of the scientific community. The important thing is to cultivate good 'morals' i.e. good manners and customs. Each member shares a responsibility for this. One way of promoting good morals could be to articulate ethical guidelines, another to have mandatory courses in research ethics for graduate students. A third way could be to have Research Ethics Committees review and monitor the work. More dialogue on ethics within the scientific community could also be of vital importance. The fourth level is represented by the government. The task of the government is to enact policies and legislation regarding research with the public good as a goal.

A key problem is how these responsibilities should be balanced. How much legislation regarding research is needed? How much responsibility should be left to the scientific community itself, to the research teams, and to individual scientists? I have already discussed this issue with regard to the choice of subject in research. Let me now briefly comment on it in more general terms.

Most important is that individual scientists and research teams assume moral responsibility for their work. Moreover, the scientific community has a responsibility to increase ethical awareness. However, there will always be a need for at least some legislation as well. Even if most scientists act according to the ethical guidelines, some do not. This is quite similar to the fact that most people do not commit murder, but some do, and therefore laws against murder are necessary. Exactly what laws should be enacted regarding research is, however, a delicate matter.

An important issue in this book regards the responsibility of scientists for the consequences of their research. I have argued that the scientists should assume such responsibility by trying to affect which applications are made of their findings. This makes it essential for the various parts of the scientific community to develop professional policies for the long term goals of their enterprises. Which consequences and applications are desirable to promote and which should be counteracted?

CHAPTER THREE

THE HUMAN GENOME PROJECT: JUSTIFICATION, PROMOTION, AND ACCESS TO RESULTS

1. SUMMARY AND APPLICATION: RESPONSIBLE GENETICS

What should responsible geneticists do? This is the key question in this and coming chapters. 'Responsible geneticists' are those who actively take moral responsibility for their research and its consequences. The term may include individuals and research teams, as well as the genetics community at large.

In answering this key question, we should keep in mind the distinction between the domain of responsibility (the crossing of a stage of research and a party involved in or affected by research), content of responsibility (the obligations and virtues), and form of responsibility (the means of implementing the normative content).

With regard to the domain of responsibility, I have suggested that scientists in general should assume responsibility for all the different stages of research. This view is also applicable to geneticists. They should take moral responsibility for the choice of subject (topic), the research process in the narrow sense (empirical and theoretical work), publication, and application. What geneticists do affects other people. Their research has consequences in the short run and in the long run. The idea that geneticists should assume responsibility for the choice of subject, the research process in the narrow sense, and publication, is uncontroversial. But that they should take responsibility for the applications of their findings is not. I have argued that scientists, including geneticists, should assume responsibility for the applications in the sense that they should try to affect applications of their findings made by other people. My main arguments are that scientists have a duty to respond to the needs and fears of the general public, because their research has a social impact, and that scientists owe this concern to society, because they have received from society education, resources, and access to the general infrastructure. 'Responding' and 'owing' correspond to the two metaphors constituting the concept of responsibility. To these arguments can be added two further arguments that are particularly applicable to genetics. Geneticists can often anticipate what possible applications might be made, and possible applications are often referred to as a justification when applying for financial grants to basic research.

In this and subsequent chapters, however, I shall not discuss these questions about the domain of responsibility any further. Instead I shall focus on the content and, to some extent, form of responsibility. I have already suggested certain general prescriptions: that scientists including geneticists should use their moral imagination to envision different consequences of their research and different ethical arguments with regard to these consequences, that they should learn from historical cases such as the eugenics movement and the recombinant-DNA debate, that they should take part in a dialogue with various parties on research and application, and finally that they should integrate ethical reflection with scientific practice, i.e. find an appropriate form of responsibility. I shall now apply these rules of thumb to different kinds of human genetics research and applications.

It should be noted, however, that the attention given to each of these general prescriptions may vary among different sections and chapters. The prescription of moral imagination will get attention throughout the book, explicitly and implicitly. It is the key idea of imaginative casuistry. Dialogue will also be central in most sections. Many of the ethical problems raised by human genetics have to do with communication of genetic knowledge and information. For instance, there is a need for dialogue in the research process between geneticists and human subjects, but also a need for dialogue between the genetics community and society at large on the consequences and applications of genetics research. Historical cases, however, will be discussed occasionally, most importantly in Chapter 6 in relation to the topic of eugenics. Integration should be viewed as a general suggestion to the geneticists to find appropriate forms of taking responsibility. For example, should the genetics community try to formulate a common professional policy on certain issues or stay content with governmental regulations? Such questions of integration will also be touched upon in a few sections.

There are, of course, many different views on what responsible geneticists should do and it is obvious that geneticists must find their own answers. They have to carry out their own interpretation, application and ranking of different ethical principles and values. However, I will propose certain suggestions about what they should reasonably take into consideration when taking responsibility. For instance, I will present different arguments in favor of and against certain kinds of human genetics research and applications. I shall also occasionally reveal my own standing on specific issues; not that my views would be particularly interesting as such, but as a contribution to dialogue.

2. DIALOGUE AND THE HUMAN GENOME PROJECT

The Human Genome Project (HGP) is of key importance to many different branches of human genetics. In this chapter, I shall discuss some of its ethical aspects. Of particular importance is dialogue. Dialogue was needed before the project started and it is also needed now when it is almost completed. Geneticists have to take part in a dialogue with other scientists, politicians, and the general public about the justification of the project. Dialogue is also an integrative part of the promotion of the project. How are other scientists, politicians, and the general public informed about the project? How is the project described and what communicative methods are used in fund-raising? Finally, dialogue is essential within the scientific community regarding the results of the project. Does the prescription of dialogue imply open access for all to the results or can restricted access in terms of gene patents be accepted in the short run?

Some Characteristics of the HGP

I shall start by giving a short description of some of the characteristics of the international Human Genome Project coordinated by HUGO. These characteristics will indicate why the HGP is such a 'hot' subject in research ethics and worthy of a thorough dialogue from an ethical point of view. In the dialogue, responsible geneticists would do well to have these characteristics in mind. What is an everyday reality to geneticists may stick out as controversial to other people.

Big science. In magnitude and costs the Human Genome Project is sometimes thought to stand in the tradition of the Manhattan Project of developing the atomic bomb and the Apollo Project of placing a man on the moon. This is clearly an overstatement but, to be sure, it is a large scale project.

Controversial. The HGP is rather controversial, but perhaps more so at the start than now when it is almost completed. Criticisms have come both from within science itself and from outside science. Some critics have questioned the scientific value of sequencing the whole genome and argued that the money should be spent on other projects instead. Others have objected to its anticipated social and medical value. If social and medical benefits are the real goals, the money could have been spent in more efficient ways, they argue.

The result of a struggle for power. Since the project is big science it is not strange that its start is characterized by a struggle for power between different interests in society. In this way the origins of the HGP illustrate perfectly the contextual interaction model proposed in Chapter 2. Clear examples are the US project and the European Community project.

The fight over the US project had to do with exactly what kind of research should be conducted and who should direct it. With regard to the first issue, it is clear that all the important scientists from whom the idea of the project originated in 1985-86–Robert Sinsheimer, Walter Gilbert, Charles DeLisi (Department of Energy), Renato Dulbecco–shared the vision of a sequencing project. However, the project was to be redefined by the National Research Council (NRC) of the National Academy of Sciences (NAS). In 1988, NRC issued a report stating the intention of mapping first and sequencing later. Sequencing should be the ultimate goal, but would require advanced technological developments. It would be wiser to start with genetic mapping and physical mapping. A pilot project for sequencing could be initiated, however. The report also stressed the importance of studying the genomes of other organisms. This revised view was embraced by the National Institutes of Health (NIH), and eventually also by the Department of Energy (DOE). Consequently, in the final joint NIH/DOE Human Genome Project three scientific goals are stated, namely physical mapping, genetic mapping, and sequencing. The second issue of controversy had to do with what agency or agencies should direct the project. As indicated, DeLisi at DOE early proposed a sequencing project. DOE wanted this kind of research as a part of its study of heritable mutations due to radiation and as a way of using its laboratories in solving nationally important problems. NIH, on the other hand, came to support the idea of a genome project somewhat later, and primarily for the reason of finding disease genes as a part of biomedical research. Eventually, DOE and NIH joined and proposed both mapping and sequencing with the justification of finding disease genes. These two agencies still have separate project directorates but cooperate rather well, for instance through certain common committees.[1]

The fight over the human genome project in the European Community had a different character. A key issue was how to characterize and justify the project. While the US struggle concerned the scientific value of sequencing, the European struggle had to do with the social value of the project. In 1988 a proposal appeared from the European Commission entitled ‘Predictive Medicine: Human Genome Analysis’. The aim of this ‘health measure’, as it was characterized, was summarized as follows: “Predictive Medicine seeks to protect individuals from the kinds of illnesses to which they are genetically most vulnerable and, where appropriate, to prevent the transmission of the genetic susceptibilities to the next generation.”[2] The genome program was expected to make Europe more competitive by slowing down the rate of increase in health costs and by making its scientific and technological base stronger. This proposal was criticized in a report by the ‘reader’ Benedikt Härlin, a member of the West German Green Party. He warned against the genome project as an enterprise in preventive medicine. In his interpretation, the proposal included eugenic elements. The application of genetic information to protect people from transmitting genetic diseases or conditions would almost always involve eugenic decisions about what is normal and abnormal, desirable and undesirable. As a result of this report there appeared a new proposal from the European Commission without regard to predictive medicine and simply called ‘Human Genome Analysis’. It was formally accepted in 1990 by the European Community Council of Ministers, authorized for three years. The genome program then became part of the ‘BIOMED’ programs.[3]

---

[1] See Cook-Deegan, 1994, for the most detailed and authoritative account of the origins of the project.
[2] Commission of the European Communities, 1988. Quoted in Kevles, 1994, p. 20.
[3] See Commission of the European Communities, 1988; Commission of the European Communities, 1989; European Community, 1989; Official Journal of the European Communities, 1990; Kevles, 1994.

Big business. Great commercial expectations are linked to the HGP. Many academic geneticists are directly involved in biotechnology companies. They serve as board members, as consultants, or even as investors. A large part of the genome research is carried out by private companies. Important examples are Celera Genomics and Incyte Pharmaceuticals.[4]
Advanced technology. The HGP uses very advanced instruments. To a large extent this technology is a product of the project itself. This is especially true of sequencing technology and data handling.
Tight links between basic research and applied research. A significant feature of the HGP are the tight links between basic research and applied research. New discoveries are rapidly put to practical use. This is true both of genetic knowledge and of the technology being developed. Knowledge of new genes is rapidly translated to clinical and other applications. Mapping and sequencing technology is transferred to and used in other kinds of research.
Teamwork. The HGP deviates distinctly from the old picture of the lonely scientist seeking the truth. Genome research is a well-organized teamwork.
International cooperation. The teamwork is carried out not only at a national level, but also at an international one. The HGP is an international project. As mentioned above, the cooperation is directed by the Human Genome Organisation (HUGO). This organization aims at coordinating the work in order to avoid overlap and making the knowledge generally available in data bases.
Competition. However, there is also competition, within nations as well as between nations, between publicly funded research teams and private companies. The cooperation is only partial. 'Gene hunting' is not a far-fetched metaphor. 'Gene wars' is another, but probably adequate only for the years preceding the start of the HGP in the US.
High scientific status. Due to its character of big science, genome research enjoys high scientific prestige. Some critics maintain that other equally important fields of investigation become second-rate, at least in the eyes of the funding agencies.
Far-reaching medical and social consequences. It already appears quite obvious that the HGP will generate knowledge and applications with great impact on society. The project can be anticipated to lead to improved human self-knowledge as well as great medical advancement. Some critics argue that some of these anticipations are unfounded. The results will not be as important as is often proclaimed. That might be, but on the whole the consequences are massive and cannot be ignored. The problem is how to avoid harmful consequences and promote good ones.
Ethical reflection as part of the project from the very beginning. Above I mentioned the comparison made by some between the HGP and the Manhattan Project. Both projects could be viewed as examples of big science. But, of course, there are differences as well. One important difference concerns ethical reflection. Right from

[4] See URL: www.celera.com, and URL: www.incyte.com.

the start, a certain percentage of the national funding was allocated to research concerning the ethical, legal and social implications of the HGP (ELSI). In the Manhattan Project no such thing was seen. Certainly, one may wonder about the motives behind this funding of ethical reflection. Critics have argued that ELSI research is intended merely to gain acceptance of risky technologies. Others argue that in the end and on the whole the funding must be seen as something positive.

Pressure on policy-making and legislation. In spite of the ELSI research, there is a tendency that the genome research is always a step ahead of the politicians and legislators. New discoveries and applications drive the development, not political decisions. This puts pressure on the decision-makers.

Each of these characteristics will be touched upon directly or indirectly in the subsequent chapters, although not in a systematic manner.

3. JUSTIFICATION OF THE HGP

One important ethical issue raised by the Human Genome Project concerns the justification of the choice of subject (topic) itself. The key question is whether it is morally responsible to carry out this kind of research. Let us take a statement from the *Declaration of Inuyama* as a starting point for a discussion of this question (CIOMS, 1995):

> The Conference agrees that efforts to map the human genome present no inherent ethical problems but are eminently worthwhile, especially as the knowledge revealed will be universally applicable to benefit human health.

This statement is interesting for a number of reasons. First, the scientific value of mapping efforts is taken for granted. No reasons are offered, despite the hard fight on this issue preceding the start of the project. However, this could possibly be explained by the exclusive reference to mapping. Sequencing is not mentioned, and it is this part of the project that has been the most controversial from a scientific point of view. Second, it is not self-evident that there are no ethical problems with the project itself. The project might have a low scientific value or a low social value. It might even have a negative social value. Third, the only explicit reason presented in favor of the project is the expected health care benefits. There may also be other possible social benefits.

It is essential that geneticists take the general debate on the HGP seriously. The HGP has been and still is rather controversial. Other scientists, politicians, and the general public may question its scientific or social value. The geneticists need to retain their trustworthiness by taking part in a public dialogue on the project and its consequences. In order to do so successfully they have to use their scientific and moral imagination in deliberating on the various arguments for and against the project.

Let us take a closer look at some of these arguments, starting with those regarding the scientific value.

Tool-Making and Its Scientific Benefits

The main argument put forward in favor of the HGP is caught in expressions like 'tool-making' and 'building a scientific infrastructure'. High solution genetic and physical maps as well as a complete sequence of DNA will be of immense value to scientists in biomedical research, in biology in general, and in many other disciplines. The HGP is a one-time effort to create tools of great expected value to science. When we know the whole sequence, it will be much easier to identify specific genes and understand their function. The project is not hypothesis-driven in the strict sense, but it gives scientists tools to be used later in hypothesis-driven research. Making such tools is quite common in other fields of science where building a telescope or an accelerator are important aspects of the work (Kingsbury, 1997).

The HGP will also generate advanced technology. At the first stage, automatic systems for test procedures, including mapping and sequencing, will be developed. At the second stage, the main objective will be information processing, handling of enormous amounts of data. These technologies will be of great use in biomedical research, health care, and other fields.

Let us take a look at some examples of how the tools and technologies created by the HGP might be used, starting with basic biomedical research. When we have complete genetic and physical maps and a complete sequence of the human genome and the genomes of the model organisms, it is argued that we have an extremely valuable basis for understanding the cell and its interaction with other cells in the body and with the environment. The results will also be of great importance for the understanding of the development of the human embryo. Moreover, the project will have tremendous importance for the understanding of monogenetic and multifactorial diseases. Monogenetic diseases are caused by mutations in a single gene, multifactorial diseases depend on both genes and environmental factors. Examples of monogenetic diseases are Huntington's chorea, Duchenne's muscular dystrophy and cystic fibrosis. The genes for these three diseases have already been found. Multifactorial diseases include, for instance, cardiac diseases, stroke, diabetes, cancers, and some mental diseases. A complete map and sequence will make it possible to identify many more of the genes involved in these monogenetic and multifactorial diseases, and at a faster pace.

The tools are also expected to lead to increased knowledge in biology in general. We will get a better understanding of the evolution of our own and other species. The HGP will not only map and sequence the human genome but also the genomes of certain carefully selected model organisms. Moreover, with the help of new technologies it may become possible to extract small quantities of DNA from extinct species, massproduce it with the PCR (Polymerase Chain Reaction) method and then analyze it. At a general level, we will get a better understanding of the similarities and differences between our species and other species, and of the similarities and differences between individuals, families and populations. The HGP will lead to increased understanding of different aspects of the developmental processes. This is so because the model organisms are in many respects ideal for such studies due to the limited size of their genomes. In particular, we will understand how different genes are regulated. With the help of the fruit fly and other organisms we will understand how human development is regulated and what has gone wrong when a child is born with developmental damages.

However, it is not only biomedical research and biology in general that are expected to benefit from the HGP but other disciplines as well. It will have important implications for the agricultural sciences. The study of genes for certain diseases, growth speed and fertility has not been analyzed by traditional genetics. Information from humans and mice can be applied to cows and pigs, and will consequently be of importance for animal breeding. In a similar way, information from model plants in genome research can be applied to plant breeding.

The understanding of the human genome will also have implications for psychology. As behavioral genetics advances we will get an understanding of the genetic basis of human behavior. We will begin to understand the complex interaction between genes, cells, organs and the environment affecting our behavior.

The HGP will also be of significance for the humanities and in this way build a bridge between the natural sciences and the humanities. Knowledge of the genome and our evolution will lead to increased human self-knowledge. For instance, the methods of the research will be of importance in archeology. The use of these methods will shed light on the origins of our species and of different populations all over the world. Of particular importance will be new knowledge of the diseases of earlier generations.

More Efficient Ways of Increasing Basic Scientific Understanding

This optimistic picture has, of course, been criticized. Let us now turn to some possible objections to the scientific value of the HGP.

One important argument focuses on the sequencing part of the project and stresses that such sequencing is of doubtful scientific interest. As we saw in the brief presentation above of the origins of the US project, sequencing was the key element from the start, although it was made only one of three objectives in the final project and perhaps even one to be started only later when appropriate technology was developed and costs were low enough. It was also the most disputed element in the whole process leading to the start of the project. Moreover, in the genome research in many European countries sequencing has been rather peripheral. Genetic linkage mapping and functional analysis have been more central. Thus, it is obvious that a sequencing project is not self-evidently of high scientific value. Critics have stressed that sequencing does not tell us what regions are interesting. Much work will be spent on sequencing 'junk', because the genes actually constitute only a small fraction of all DNA. The funding and the energy of the scientists could be used in better ways. We can discover genes without knowing the whole sequence (Rosenberg, 1996).

Another argument focuses on the 'big science' character of the HGP and maintains that the high costs are not justified. With respect to the US project, it has been argued that it would be more efficient to let individual research teams apply for federal funding as in ordinary research and study areas of their own choice that they judge to be of particular interest. This would yield more interesting results more efficiently and at a lower cost to society.

'Geneticization' of Biomedicine and Biology

It has also been objected that the HGP promotes a 'geneticization' of biomedicine and biology. The focus on genes draws the attention from other important scientific issues. One example is the issue of the developmental processes. These processes are at least as important as the genes for the understanding of organisms, including humans. More funds should be allocated to this kind of research instead of mapping and sequencing the genome. Knowledge of the human genome is not the 'Holy Grail' that makes us understand everything, as some of its proponents tend to see it. The molecular genetics approach to biology has been very successful but there is a risk that it will impede other at least equally fruitful approaches. In particular, there is a risk that the importance of environmental factors will be ignored (Lewontin, 1992, pp. 59-83).

The Goal of Biomedicine: Cure or General Knowledge?

Critics have also focused on the general view of biomedicine implied in the HGP. The central aim seems to be to increase knowledge, not to help people. Certainly, the helping aspect is part of the proposed justification of the project but in real work the emphasis is on gaining new knowledge, not on finding efficient cures. Scientific value in biomedical research should be related to the cure of diseases, not to general knowledge about human biological phenomena. If the goal truly is cure, the money should reasonably be allocated to more clinically urgent fields of research.

'Geneticization' of Health Care and Society

Moreover, it is argued that the HGP exacerbates, if not initiates, the ongoing 'geneticization' of health care and society.[5] In health care, the genetic background of many diseases is more and more recognized. Much hope is also linked to treatments based on gene technology, but there is a risk that this will draw the attention from nongenetic causes of disease and from nongenetic treatments. For example, this geneticization might divert the attention from the social causes of ill health. There is also a risk that gene therapy and genetically tailor-made drugs will be perceived as miracle cures, solutions to all health problems.

In society at large, the gene has become a cultural icon. The expression "it is in the genes" is a part of everyday language. This means that there is a risk that reductionistic ("Humans are their DNA") and deterministic ("DNA is destiny") views on humans will become prevalent in society. In their selling of the HGP even leading geneticists use this kind of language (see pp. 106-109).

This focus on genes might have serious social consequences in terms of discrimination and stigmatization. There is a risk that genes will be viewed as causes to almost every aspect of human life, and with this follows a risk that things will be understood as immutable. This might, in turn, be used as a way of legitimizing inequalities in society. In other words, there is a risk that the genes will acquire an ideological function. On the basis of genetic determinism it might be argued that nothing can be done to get rid of social inequalities. Poor people are poor because of low intelligence and low intelligence is to a very large extent determined by the genes. Low intelligence "is in the genes". Consequently, the situation of poor people cannot be improved by social means. This becomes particularly serious when low intelligence is supposed to be linked to certain ethnic groups.[6] In a similar way, male dominance in society might be legitimized. On the basis of genetic determinism it might be argued that male dominance over females "is in the genes", and if it is determined by the genes, it cannot be changed by social effort.[7] Other examples are discrimination and stigmatization of disabled people at the workplace or in insurance (Asch, 1996). There is also a risk that social factors will be ignored in explaining criminality, and consequently also in taking measures against it.

Discussion

How should these arguments for and against the scientific value of the HGP be assessed? Responsible geneticists should, of course, make their own assessment. However, as a contribution to dialogue, I shall make a few suggestions.

---

[5] See, for instance, Nelkin, Lindee, 1995; Keller, 1995; Lewontin, 1992.

[6] This view is defended in Herrnstein, Murray, 1994, *The Bell Curve*. This book will be discussed on pp. 111-114, 227, 234.

[7] Karen Lebacqz criticizes this view in Lebacqz, 1999.

The mapping part seems rather unproblematic. The sequencing of the entire genome including all the 'junk DNA' appeared at least initially, however, more disputable. There seemed to be more efficient and less costly means of obtaining basic scientific understanding. It could be argued that more effort should be put on the identification of disease genes and on function analysis. However, now when the sequencing is almost completed, it no longer seems as problematic. The sequencing has been carried out much faster than anyone expected, and to a large extent by private companies.

But if sequencing, not without reason, at least initially appeared to be of doubtful scientific interest, why did the community of molecular biologists stress its importance so persistently? The philosopher Alexander Rosenberg suspects that the real motive behind the sequencing project was a desire for technological spin-offs: research methods, computer programs, automatic systems for sequencing etc (Rosenberg, 1996, p. 12). If this suspicion is true, it could be argued that the geneticists for ethical reasons should have openly admitted this motive.

Another problem raised by the HGP concerns its implied view on the goal of biomedical research. Is its goal cure or general knowledge? Should the scientific goal of biomedical research be cure or general knowledge? However, this is a wrong way of phrasing the question. There need not be an either/or, neither generally nor with regard to the HGP. However, it is important that the basic science character of the HGP, not just its expected health benefits, be openly acknowledged. Certainly, it might be tactical to stress the health benefits in order to sell the project to the general public and the politicians, but it would be more morally responsible to play with open cards and present the project as primarily a basic science project.

What about the objections stressing that the HGP might exacerbate the 'geneticization' of science, health care and society? They seem to carry some conviction. There is a risk that the focus on genes might draw the attention from, for instance, the developmental processes and from the significance of environmental factors. This risk for a geneticization of science has two aspects. The first is economic and concerns the amount of resources allocated to different kinds of research. If the resources were to be taken from other fields of research, this would be a serious problem. However, if they constitute new funding, then the problem diminishes, although the comparatively greater amount of resources allocated to the HGP may still be problematic in itself. The second aspect of the risk concerns the content of scientific results. The focus on genes might lead to a genetic determinism that downplays the importance of environmental factors in biological processes. It is vital that scientists working on the HGP, as a part of their scientific responsibility, fight genetic determinism and stress the importance of nongenetic factors. Geneticization of science should be balanced with other perspectives. The complexity of biological processes should always be stressed.

Regarding the risk for geneticization of health care and society at large, it is important that the geneticists working on the HGP take responsibility for fighting such tendencies and their consequences. I agree that a geneticization of belief systems and social life is dangerous.

The objections focusing on geneticization are, in my opinion, not strong enough to judge the project unethical, but point at risks that geneticists should be aware of and fight as part of their moral responsibility.

In conclusion, all the objections have certain valid points although they are not strong enough to justify the view that the project is ethically unacceptable. However, they say something important about what geneticists working on the HGP should do as part of their moral and social responsibility.

Social Value?

The potential social benefits of the HGP are often referred to in its justification. It is thought that applications made on the basis of knowledge obtained through the project will benefit humankind in several ways. This is clearly seen in the *Declaration of Inuyama* which in particular stresses the health benefits. Examples of applications in health care are genetic testing and screening, genetically tailor-made drugs, and gene therapy. Note that some of these technological developments were to some extent initiated even before the HGP started. However, the information obtained through the HGP will be of great importance for further developments. Therefore, they can be viewed to a large extent as applications of this project.

If applications are often referred to in the justification of the project, they are also often referred to in the criticism. Actually, the most important objections to the project concern certain applications and their social consequences. Some critics see these objections as being so serious that they conclude that the choice of subject itself is unacceptable. These fears have to be taken seriously by responsible geneticists. They need to balance the social pros and cons of the project.

Health Benefits and Other Social Benefits

The proponents of the HGP stress that as a result of the project more efficient and reliable methods for genetic testing can be expected to be developed. There are many different types of tests, including carrier identification, prenatal testing, preimplantation genetic diagnosis (genetic diagnosis on fertilized eggs in the process of *in vitro* fertilization), newborn screening, testing on late-onset disorders, and DNA fingerprinting. Tests used to identify whether a person has a genetic condition or a disease may be requested by individuals if there is a family history of a specific disease, if they show symptoms of a genetic disease, or if they are concerned about passing on a genetic disorder to their children. Screening might be of importance as a preventive public health measure. At the workplace it might be possible to identify people with an increased risk of being hurt by exposure to toxic substances. DNA fingerprinting may be used in legal cases involving paternity and in criminal investigations. It may also be used in wars, disasters and accidents to identify the dead.

The findings of the project are also expected to improve the possibilities for genetically engineered drug design. New drugs can be developed using gene technology. Even before the HGP started human insulin and human growth hormone were produced by transgenic animals. We will see much more of such developments as we acquire more knowledge of the genome through the HGP. Moreover, when genetic tests are developed on the basis of the new knowledge of the human genome, drugs can be tailor-made for individual expressions of diseases. In this way side-effects of drugs may be reduced and efficiency increased. This 'pharmacogenomics' has a very great potential.

Because of the shortage of organs for transplantation, there will also be a demand for genetically engineered animal organs for human transplants. The work on xenotransplantation will be informed by the new genetic knowledge obtained through the HGP. By transfer of human genes to the animals or by 'knock-out' of animal genes (i.e., removal of the genes or making them inactive), it may be possible to avoid rejection of the transplanted organs.

The knowledge generated by the human genome research is also anticipated to be of great value for the development of gene therapy. Many scientists hold that gene therapy has an extraordinary potential in the long term for the management and correction of human diseases. One may distinguish between gene therapy on somatic cells and gene therapy on the germline. The changes made in the latter kind of therapy are inherited. It is therefore very controversial and many of those in favor of gene therapy on somatic cells are against it. Presently, only somatic gene therapy research is being carried out.

Finally, some of these kinds of applications–primarily genetic testing and genetically engineered drug design–may have a great potential not only for health care, but also for industrial growth. They are expected to make valuable contributions to national welfare and making nations more competitive.

More Efficient Ways of Improving Health?

Reference to health benefits is often made in the justification of the HGP. However, if we turn to the criticism of the social value of the HGP, we find the objection that if the goal is improvement in health there may be more effective strategies than mapping and sequencing the entire human genome (Paul, 1991). Even if we focus on genetic medicine, it would be better to try to directly identify disease genes and study their function. Moreover, there might be other medical measures of greater importance. Finally, if we spent more resources on caring, prevention and protecting patients, our need for high tech solutions would be dramatically reduced.

There is also an even more radical objection, stressing that instead of spending national funds on genome research they could be used in medical and other programs for helping poor people in the third world. This argument can also be used against other forms of high technology in medicine. It should be distinguished from the argument mentioned above, which points out that there could be a better scientific use of the money. Here it is a matter of a better social, not better scientific, use.

Objections to the Applications

Other objections focus on the applications as such. Some arguments state categorically that a particular application is morally unacceptable because it violates a person's integrity. Other arguments judge an application morally unacceptable depending on its consequences, for instance, social consequences.

I will discuss extensively arguments for and against different applications in Chapters 4, 5 and 6. As an illustration, let me give a couple of examples of objections. The first is an objection to genetic testing. It is a much discussed objection that stresses that genetic testing poses social risks such as discrimination in insurance. Persons with the 'wrong' genotype might have to pay extremely high premiums for insurance. The second example is an argument against gene therapy. It is pointed out that if we permit gene therapy we might end up in a slippery slope. Even if we start with the intention to do gene therapy only on very serious diseases, there is a risk that we will see a development in society in which parents try to design perfect babies with genetic enhancement techniques, and this is highly undesirable.

Discussion

Let me first comment on the argument that there are more efficient ways of improving health than starting a gene mapping and sequencing project. This might be true in the short run. Genetic medicine will probably be of great value in the long term. But at present there are many other things to do that will, faster and at a lower cost, improve health nationally as well as globally. The vast amount of money spent on the HGP should perhaps instead be spent on other kinds of medical research, or on caring and prevention. It should be noted, however, that this objection could be directed not only against the HGP but also against most other kinds of high technology in medicine.

The objection is not, however, strong enough to judge the HGP as ethically unacceptable. But it is an important memento. High technology is not the solution to all health problems. We need both high technology and good personal care. The balancing is a delicate matter.

But what about the possible applications themselves? How should they be evaluated? It is not difficult to recognize the social value of many of the anticipated applications. The problem is how the objections should be assessed. As indicated, I will discuss these issues extensively in coming chapters. Let me just anticipate one of the conclusions and state that, in my opinion, the objections are not decisive enough to suggest that the HGP is socially irresponsible. What is suggested by the valid points of the objections is that responsible geneticists should try to take measures to avoid certain bad consequences of its applications. They can, at least to some extent, try to influence the uses of scientific results in a desirable direction. A moderate optimism concerning the possibility to affect future technological development including the development of gene technology is to be preferred. Our society can, at least to some extent, avoid bad consequences if we become aware of them in time. It would be morally responsible to carry out the HGP only if it could be expected that the benefits clearly outweigh the risks. I believe that this is the case.

In sum, there are many good potential applications of the HGP, although there are also some risks. The geneticists, and first and foremost the genetics community at large, have a responsibility to fight bad applications and promote good ones.

Conclusion: Balancing Scientific and Social Value

On balance, it seems that a good case can be made in favor of the HGP. Most of the objections to the project have some valid points, but they are not serious enough to render the project ethically unacceptable. On the other hand, the tools made available through the project seem increasingly promising, scientifically as well as socially. What is vital now when the project is almost completed is that much effort be put into more empirical research and moral reflection on the ethical, legal and social implications of various applications of the findings. How should the implications be handled?

4. PROMOTION OF THE HGP

Justification of the Human Genome Project is one thing, promotion quite another. Already from our brief look at the struggle for power preceding the start of the project in the US and the European Community, the importance of promotion is obvious. However, there are many different ways of promoting the project, and which ones are actually chosen depend on the specific social and political context in each country. Moreover, the methods have ethical implications. The questions arise: What methods for the promotion of the HGP are morally acceptable? What does a responsible dialogue with the general public and the politicians imply?

One common method is to stress the medical benefits of the HGP. We can see this, for instance, in the *Declaration of Inuyama*: "... the knowledge revealed will be universally applicable to benefit human health" (CIOMS, 1995). Such applications are, of course, something good. Moral responsibility demands, however, that the benefits are not overestimated in order to 'sell' the project.

Another method is the rhetorical use of metaphors denoting the importance of genes for diseases and nondisease characteristics. As we will see below, these metaphors often tend towards genetic determinism, reductionism, and essentialism. There are some clear examples of this from leading scientists in the US. While acknowledging the importance of genes, I will argue that the promotional use of such deterministic, reductionistic and essentialist metaphors is neither scientifically nor ethically acceptable.

A contrasting method of promotion is to see to it that human genome research appears as a cautious, responsible and trustworthy enterprise. This can be achieved by clear statements against certain far-reaching applications of the findings such as germline gene therapy, and against overinterpretations of the results such as genetic determinism. It is my impression that in, for example my own country, Sweden, this is not an uncommon strategy of leading geneticists. Another example can be seen in the debate preceding the European Community project referred to earlier. Due to the discipline's eugenic past, it seemed important, not least in Germany, to present the project as noneugenic. However, even this 'trusteeship' method might be problematic although it is on the whole very sympathetic. If this approach is merely a matter of 'smart' strategy, it could easily be criticized later when more questionable applications nevertheless appear in spite of earlier declarations.

Finally, we have the method of funding ELSI research. As already indicated, it is obvious that such research consciously or unconsciously might have a legitimizing function. If this funding is intended merely to gain acceptance of risky technologies, it is, of course, not morally acceptable. However, if it is not so intended, and the scientists and ethicists are aware of this possible function and keep a sound critical attitude, then the funding of ELSI research in the end and on the whole must be seen as positive (cf. Juengst, 1996).

Responsible Dialogue?

Below I shall discuss more extensively one of these methods, namely the rhetoric use of promotional metaphors. One reason for this is that this method is perhaps the most controversial. Another reason is to illustrate and reflect on the role of metaphors in the framing of moral situations, an important aspect of imaginative casuistry. I shall also broaden the perspective and turn to the more general question of fighting misrepresentation of genetics. As an example I choose the lack of criticism of the book *The Bell Curve* from the most important genetics organizations in the US. This avoidance of criticizing the book may be understood as a part of the promotion of the HGP. By criticizing the book's exaggeration of the significance of genes, there would be a risk that the funding of the project might be reduced. This avoidance, however, seems to be morally irresponsible.

The Use of Promotional Metaphors

Many different metaphors have been used in promoting modern genetics and the Human Genome Project. One scholar who has particularly studied the promotional metaphors used in relation to the Human Genome Project is Dorothy Nelkin. She has also–with Susan Lindee–studied the overall use of genetic concepts in modern culture. She has found that the gene has become a very important cultural icon of our age (Nelkin 1994; Nelkin, Lindee, 1995). However, I will not discuss her work here. Instead I will make my own analysis of some important rhetoric metaphors used in favor of genetics and the Human Genome Project.

One important metaphor is 'The human genome is the Holy Grail' metaphor. Walter Gilbert, Nobel laureate and one of the early promoters of the HGP, has used this metaphor. He argues that the knowledge of the human genome will be the Holy Grail of modern biology. The Holy Grail was thought to bring immense knowledge to its discoverer. Gilbert's point is that when modern biology gets hold on the genome it will be an invaluable source for understanding human beings. The sequence of the human genome is "the ultimate answer to the commandment 'Know thyself'".[8] Gilbert has also used another metaphor that points in the same direction: 'A human being is a compact disk'. He introduced his public lectures on gene sequencing by pulling a compact disk from his pocket and saying: "This is you".[9] Taken at face value, these metaphors suggest that a human being can be reduced to the genome and that the essence of a human being is the genome.

Another kind of metaphor has been used by James Watson, Nobel laureate and first director of the Human Genome Project in the US. He asserts that "in large measure, our fate is in our genes"; not, as people used to believe, in the stars.[10] Very similar to this 'Genes are fate' or 'DNA is destiny' metaphor is the 'The human genome is a crystal ball' metaphor. According to this metaphor, geneticists are like fortune-tellers who have substituted genetic tests for the crystal ball. Other metaphors that point in the same direction are 'The human genome is a computer program' metaphor and 'the human genome is a blueprint' metaphor. These metaphors do all suggest that human destiny is pregiven, predetermined and predictable.

With this in mind, it seems obvious that many leading persons in genome research use metaphors in their public talks that come very close to genetic reductionism, genetic essentialism and genetic determinism. But how should these concepts be understood? To what extent do reductionistic, essentialist, and deterministic assumptions actually lie at the foundation of human genome research? Moreover, how can the fact that leading geneticists use this kind of language be

---

[8] Jaroff L., 1989. "The Gene Hunt". *Time* 20 March, pp. 62-67; quoted in Kaye, 1997, p. 184. Bishop, Waldholz, 1990, p. 218.

[9] See Nelkin, Lindee, 1995, p. 7.

[10] Quoted in Wright, 1990.

explained? These are important questions raised by the use of these promotional metaphors.

First, by classifying these metaphors as reductionistic, essentialist and deterministic, what do we mean? With respect to genetic reductionism, one should make a distinction between ontological reductionism and methodological reductionism. In the context of human biology, methodological reductionism is the view that the proper way of studying human beings is to focus on their compositional units, i.e., ultimately their genes. Ontological reductionism goes one step further, not prescribing a methodology but raising claims about human nature. According to genetic ontological reductionism, humans are 'no more than' the genes–what human beings are is specified by the genes. Thus, ontological reductionism as applied to genetics and human beings coincide with genetic essentialism. The genes are the essence of human beings. 'Humans are their DNA'. Our nature is in the genes.

Furthermore, genetic determinism is the view that 'DNA is destiny'. The phenotypic properties of human beings are caused by the genes. That these properties are caused by the genes means that the content of the properties completely or at least to a very large extent are specified by the genes.

None of these assumptions is actually necessary for human genome research, although some geneticists probably accept one or more of them. It is possible to do this research and at the same time reject them all. None has as yet been demonstrated to be valid.

Genetic methodological reductionism, no doubt, has been very successful to date. But this does not mean that it is the only acceptable methodology in biology. A focus on other levels of organization in organisms might also be legitimate. Genes only function within cells and cells are parts of tissue and organs. There might emerge properties at these higher levels of organization that cannot be reduced to genes. The interaction between genes, the cellular environment, and other aspects of the environment should not be neglected. Moreover, it is important to note that methodological reductionism does not in itself support ontological reductionism. One may defend genetic methodological reductionism without necessarily accepting genetic ontological reductionism and genetic essentialism. Methodology is one thing, human nature quite another.

An objection to genetic ontological reductionism and genetic essentialism is that they do not explicitly recognize the fact that genes can do nothing without cells. Moreover, genes can only specify amino acid sequences of proteins. The most a complete sequence of the genes can tell us is what proteins a human organism can make. Humans are more than their genes. Genes are very important but they are only part of what it is to be a human being. There might emerge properties at these higher levels of organization that cannot be reduced to genes.

To genetic determinism it can be objected that we have no evidence that phenotypic properties of human beings are completely or to a very large extent caused by genes. Genes may be necessary determinants but they are clearly not sufficient. Phenotypic properties are the outcome of a complex interaction between genes and the environment. Genetic determinism cannot as a general view be justified on the basis of present scientific findings.

Since the assumptions of genetic ontological reductionism and genetic determinism are neither necessary for carrying out genetics research nor demonstrated to be true, it is intellectually and morally irresponsible by leading geneticists and others to talk as if the one or the other is valid. Moreover, to present genetic ontological reductionism or genetic determinism as valid may have serious consequences. First, it might divert the attention from the social causes of poverty and ill health. Second, it presents an unscientific view of human beings in scientific clothes. The reduction of human beings to DNA goes far beyond science. Third, these kinds of metaphors might influence the public to an extent that may have serious social consequences. Society might become 'geneticized' with the effect that people are stigmatized or discriminated because of their genotypes.

But why do leading geneticists use this kind of language? One possible explanation is that they actually believe in genetic reductionism and genetic determinism, in spite of the fact that genetics cannot support this view, at least not for the moment. In their opinion, these ideas may be the anticipated results of genome research. In other contexts, however, they have stressed the complexity of gene/environment interaction. Therefore, a more probable explanation is that these leading geneticists are 'selling' their scientific enterprise. They want their field, human genome research, to get more funding.

However, it should be pointed out that not all of those who are positive to the Human Genome Project use reductionistic, essentialist and deterministic metaphors. The *Declaration of Inuyama* criticizes genetic reductionism and genetic essentialism in stressing the inadequacy of a view that "... diminishes human beings by reducing them to mere base pairs of deoxiribonucleic acid (DNA)" (CIOMS, 1995). A clear example of a nondeterministic metaphor can be found in a suggestion by Eric Juengst, the first Chief of the ELSI Branch at the NIH. As an alternative to deterministic metaphors, he suggests the 'Genetic testing is a weather forecast' metaphor. According to Juengst, this kind of metaphor is in many cases more accurate. It takes seriously the contingencies of the human condition and avoids genetic determinism. In the case of cancer, we may predict on the basis of a positive test result that a person is likely to face a deadly oncologic storm, but the storm has not yet hit him and will perhaps not hit him at all. What eventually happens will depend on other aspects of his genome and the environment (Juengst, 1995a).

It is interesting to note that now when the HGP is almost completed, many prominent geneticists explicitly reject genetic determinism and reductionism. In February 2001, J. Craig Venter et al. published Celera Genomics' draft of the human genome. They commented their results in *Science*. At the end of the paper they wrote (Venter et al., 2001):

> There are two fallacies to be avoided: determinism, the idea that all characteristics of the person are "hard-wired" by the genome; and reductionism, the view that with complete knowledge of the human genome sequence, it is only a matter of time before our understanding of gene functions and interactions will provide a complete causal description of human variability.

For an external observer, it seems obvious that they wrote this not only because they are convinced that the genetic determinism and reductionism are wrong, but also to show the general public that geneticists are 'good guys' and that one does not need to fear their results.

In conclusion, geneticists, health care professionals, and bioethicists should be very careful when they choose their metaphors. By using inaccurate metaphors we may create problems for ourselves as well as the public. Metaphors have great rhetoric power. We should take them seriously. They affect how we think and act. Different metaphors have different implications. When we use promotional metaphors in the context of the Human Genome Project, we should avoid reductionistic, essentialist and deterministic metaphors. They are misleading, since many people will probably understand them as scientific truth. Moral responsibility in genetics demands that we use adequate metaphors. These might in the short run not be as efficient for fund-raising purposes but they are more honest and probably more efficient in the long run. In order to be morally responsible, our promotional metaphors should be scientifically adequate. They should not go beyond what is scientifically established at the time, and if they go beyond science, this should be explicitly admitted and not be presented as science.

Dialogue: Fighting Misrepresentation of Genetics

I have argued that an important aspect of the moral responsibility of scientists is to fight misrepresentation of scientific findings. Such a responsibility is important not least in genetics. One of the social consequences of the HGP is a growing awareness in society of the importance of genes. This increases the risk for overinterpretation of the scientific results. Such overinterpretations may be made, for instance, by journalists and politicians. The *Declaration of Inuyama* explicitly stresses the importance of correcting reductionistic accounts by educating the public (CIOMS, 1995):

> Public concern about the growth of genetic knowledge stems in part from the misconception that while the knowledge reveals an essential aspect of humanness it also diminishes human beings by reducing them to mere base pairs of deoxyribonucleic acid (DNA). This misconception can be corrected by education of the public and open discussion, which should reassure the public that plans for the medical use of genetic findings and techniques will be made openly and responsibly.

However, the *Declaration* ignores the fact, described in the previous section, that misrepresentations of genetic findings may also be made by the geneticists themselves. It also ignores that they may be made by scholars in other disciplines.

Misrepresentations are particularly serious when they are made for political or ideological reasons. An important aspect of assuming moral responsibility for the uses of the results of genome research is to fight this kind of misrepresentation.

The Bell Curve Case

The opinion that geneticists have a responsibility to fight misrepresentations of scientific findings is an uncontroversial demand. However, in recent history there are examples of neglect in this respect. I am thinking of the neglect of major genetics organizations in the US to fight the misrepresentation of genetics occuring in the widely spread and widely discussed book *The Bell Curve*. None of the authors are geneticists. The now deceased Richard Herrnstein was a scholar in psychology. Charles Murray is a scholar in political science. The Joint NIH/DOE ELSI Working Group released a statement against the book (ELSI, 1996) but organizations such as the American Society of Human Genetics, The American Society of Medical Genetics, and The Genetics Society of America, did not, despite the fact that the ELSI Working Group encouraged them to at least make a comment (Beckwith, 1997).[11] Below, I will briefly analyze the argument of the book, making its assumptions explicit. I will then present some arguments for the assertion that *The Bell Curve* constitutes a misrepresentation of results in genetics research. I shall analyze the book only to the extent necessary to substantiate my point that it represents an overinterpretation and that this misrepresentation should have been fought.

The main argument of *The Bell Curve* can be summarized as follows. The authors maintain that intelligence, measured with IQ tests, varies a great deal between different ethnic groups. They assert that IQ is to a very large extent inheritable and that differences in IQ can be explained to a very large extent by genetic factors. Moreover, they argue that social problems are caused by low IQ and suggest as a long term solution to these problems a reduction of social welfare and education. Social welfare reduces the average intelligence of the population, since it leads to poor women in certain ethnic groups giving birth to more children. Education will not pay off, since IQ is so strongly genetically determined that it cannot be affected by pedagogical means. Finally, they argue that we must somehow

---

[11] However, as the Chair of the Working Group, Lori Andrews, stresses, the NHGRI administration to whom the Working Group reported failed to make the statement public for months (Andrews, 2000, p. 196).

find a "valued place in society" even for the "dull and ignorant", since we cannot possibly improve their situation (Herrnstein, Murray, 1994, pp. 19-24, 527-552).

The argument in *The Bell Curve* rests on certain assumptions (cf. Gould, 1996, pp. 367-390). The first assumption is that there is an ability called general intelligence (g), that this ability can be measured by IQ tests and assigned a particular value (the IQ), and that individuals can be ranked linearly according to their IQ. The assumption is, as the authors admit, challenged by other schools of psychology, for instance the view that intelligence is not just one general factor but many different factors (the theory of multiple intelligences). But, and this is actually very surprising, they do not provide any justification whatsoever for this assumption, despite the fact that the book includes several hundred pages.

An implication of the idea that individuals can be ranked linearly with respect to their IQ's is that ethnic groups can also be ranked in a linear order due to their average IQ numbers. On this basis, the authors maintain that IQ tests show certain differences between the average IQ of different ethnic groups. Caucasian Americans have an average IQ of approximately 100, Asian Americans slightly more, and African Americans 85. Here, it might be appropriate to explain the title of the book. Bell curves are the same as what is otherwise called normal distributions. They occur when variation is distributed randomly around the mean. According to the authors, the bell curve of black people in America refers to the distribution of IQ's around the mean of IQ 85, while the bell curve of white people refers to the distribution of IQ's around the mean of approximately 100. There is no reason to doubt that these numbers are correct but how they should be interpreted and explained is something different. A common objection is that IQ tests are culturally colored, that they are biased in favor of white people. But Herrnstein and Murray maintain that it has been possible to construct IQ tests without such a bias (Herrnstein, Murray, 1994, p. 23). It is obvious, however, that even if the tests are unbiased and the ethnic differences in scores are correct, it is by no means clear that the differences cannot be explained by reference to social factors.

The second assumption is that IQ is to a very large extent genetically determined. To exactly what extent is not quite clear. Explicitly Herrnstein and Murray talk about 60% but occasionally one gets the impression that they tacitly mean 80-100%. For their argument to have any possibility to work they must assume the latter. This is so because of their third assumption, namely that IQ is immutable. In the opinion of Herrnstein and Murray, IQ cannot, other than very marginally, be affected by social and educational efforts, because it is to a very large extent genetically determined. To this it can be objected that a heritability of 60% certainly means that the genes are more important than the environment in determining intelligence, but it also means that the environment plays a important role. To draw the conclusion that IQ is almost immutable on the basis of the assumption that it is only 60% heritable is clearly mistaken, and this is a critical mistake because the authors propose a social program based on this conclusion.

Moreover, if, in order to give their argument at least some possibility to work, they tacitly assume that the percentage is much higher, then they have no empirical support whatsoever for this assumption. In this case, they go far beyond what genetic and behavioral studies indicate. Thus, if they assume a heritability of 60%, they have some empirical support but no reason whatsoever to conclude that IQ is almost immutable. If they assume a heritability of 80-100%, they completely lack empirical support but have at least some possibility to conclude that IQ is effectively immutable.

However, even the third assumption is open to certain objections. First of all, it is evident that even if IQ is to a very large extent genetically determined, this does not in itself mean that it is impossible to change. This can be shown by a simple medical example; a monogenetic condition like phenylketonuria can be prevented by changing the diet (Steen, 1996, pp. 15-18).

Moreover, it seems quite possible to explain the differences in IQ between different ethnic groups by differences in social environment. In support of this suggestion, I would like to refer to (1) historical changes in average IQ within ethnic groups, and (2) IQ studies on poor African American children adopted into white middle-class households.

It is a fact that since the 1930s the average scores in IQ tests of white Americans have risen about 14 points. The scores of black Americans have also risen but not as much. This indicates that IQ is not rigidly fixed by the genes (Kevles, 1995, p. xi). This is part of a general phenomenon called 'Flynn's effect' i.e. the observation that IQ's have been steadily growing within various ethnic groups since the turn of the century (Flynn, 1980).[12] This observation by James R. Flynn has been widely discussed. One recent example is *The Rising Curve* edited by Ulric Neisser who calls Flynn's achievement "transforming work" (Neisser, 1998). Note the title of the book which obviously is intended to contrast *The Bell Curve*.

Moreover, the influence of social environment on racial differences in IQ is also indicated, for example, by the study of Sandra Scarr and Richard Weinberg on 99 poor black children adopted into white middle-class households. The average IQ of these adopted children was 106, which was higher not only than the average black IQ but also than the average white IQ. Scarr and Weinberg estimated that being raised in a relatively privileged social environment contributed about 16 points to the IQ scores (Scarr, Weinberg, 1976).[13] Interestingly, this is just about the size of the difference between average IQ scores for whites and blacks, as stated by Herrnstein and Murray.

In conclusion, it is obvious that Herrnstein and Murray misrepresent science. They go far beyond what genetic and behavioral studies can justify. In asserting that IQ to a very large extent is genetically determined and in maintaining that IQ is a very central trait of character that determines the socio-economic standard of people,

[12] See also Holloway, 1999.

[13] See also Hamer, Copeland, 2000, pp. 224-225.

they not only adhere to a kind of genetic determinism, they also in effect accept a kind of genetic reductionism. What people are is in their genes. Humans are reduced to their DNA. The views of Herrnstein and Murray are based on assumptions of genetic ontological reductionism and genetic determinism that are not scientifically warranted, at least not at the moment. Certainly, there is an important genetic background to intelligence, indicated by twin studies.[14] There might even be genetic differences between different ethnic groups with regard to intelligence. But Herrnstein and Murray go beyond present-day genetic results. Most important, it is not at all clear that it is not possible to modify intelligence by social means.

However, as stated above, my reason for discussing *The Bell Curve* case is not purely scientific. It concerns moral responsibility. For ethical reasons the authors should not make such overinterpretations of science. Moreover, as I have also stated above, the genetics organizations in not disclaiming the book, did not take their moral responsibility. They should have criticized the book as did the ELSI Working Group, although not necessarily in the same way. Of course, one can speculate about the reasons why they did not do so. Perhaps they agreed with Herrnstein and Murray, or perhaps they did not want to reduce the expectations of politicians and the public regarding genetics research. In that case, the funding of genomics might be reduced. They may have remained silent for promotional reasons. But to this, one could reply that in the long run a better way of promoting human genetics research would be to criticize overinterpretations of the results. Only in this way can the trust of the general public be retained.

5. GENETIC KNOWLEDGE: OPEN ACCESS OR PRIVATE PATENTS?

A major problem in human genome research concerns who should have access to the scientific findings. From the perspective of the geneticists, this problem of common access versus private patents is probably one of the most important issues. It is a central problem of social responsibility and professional policy. Partly, it can be described as a problem of dialogue. Does the prescription of dialogue imply open access for all to the results or can restricted access in terms of gene patents be accepted in the short run? How should responsible geneticists proceed?

In line with what can be expected from the perspective imaginative casuistry, the problem is to a large extent a matter of choice of prototypical case. One possible prototypical case is the long academic tradition of open access. If geneticists take this tradition as prototype, they will likely stress the importance of open access to the results of human genome research and be critical to gene patenting. An alternative prototypical case is the rather long industrial tradition of patenting. If geneticists take this tradition of patenting as prototype, they will likely be in favor of gene patenting.

---

[14] See, for instance, Pedersen, Plomin, Nesselroade, 1989, and Scarr, 1992.

During the last decades a new kind of university-based science has been emerging, which takes the industrial tradition of patenting as prototype. It is characterized by 'patent first, publish later'. This development can be seen, for instance, in pharmaceutical and technological research. What is happening now is that we see the same development in genetics research. University professors are funded by or directly involved in biotechnological companies. This holds for basic research in order to find disease genes as well as for applied research in order to find cures. Gene patenting is very important for these companies.

However, the fact that gene patenting works against the tradition of shared knowledge in scientific discovery makes it controversial. Among the general public, it is obvious that some people think of it in terms of patenting life and finds it repugnant because 'no one can own life'. But gene patenting is disputed also among scientists. Some criticize it for conflicting with the prototype of academic openness. Others think that gene patenting should not be accepted for more pragmatic reasons. Science proceeds more slowly if publication is postponed. Geneticists in favor of gene patenting, on the other hand, often stress that cures for diseases will come faster with financial inputs from biotechnological companies ('no patents, no cure').

A fundamental conceptual problem in this discussion is, of course, what is actually meant by 'gene patents'. What is a patent? What exactly are we thinking of when we talk about 'gene patents'? In order to have a well-informed discussion, it is important to avoid certain misunderstandings of the nature of patents in general and of gene patents in particular. This is essential in the dialogue between geneticists and the general public. The central problem of the section concerns, however, who should have the access to the findings of the HGP, all aspects considered, academic as well as commercial.

I shall start by briefly discussing the commercialization of genetics. This provides a realistic framing to our discussion of gene patents. The major part of the remainder of the chapter consists of an analysis and discussion of various arguments for and against 'gene patenting'.

Commercialization of Genetics

The commercial aspects of basic and applied genetics research are becoming more and more obvious, and this is of great relevance to the issue of social responsibility in human genome research. Many new biotechnology companies are searching for disease genes. Genetic tests are developed for the market, and there is a growing awareness of the possibilities for genetically tailor-made drugs. The financial stakes are enormous. The consequence is a strong drive to patent genes that are discovered.

This brings us to problems regarding the relation between scientists and the biotechnology industry. Many academic geneticists are linked to biotech companies. Some of them are board members, others function as consultants, and, of course, many geneticists have invested in such companies. Is this ethically acceptable? How do such connections influence academic research? How should the responsibility towards the academic institution be balanced against the responsibility towards the company? How should the responsibility towards the company be balanced against the responsibility towards the general public? These are all very important questions.

The recent commercialization of genetics is very clearly illustrated by the career of Walter Gilbert. After a very successful academic career crowned by the appointment as American Cancer Society Professor of Molecular Biology at Harvard and the Nobel Prize in chemistry, he became involved in the founding of the Biogen company at the end of the 1970s. He assumed positions as acting president, chairman of the scientific board, and finally cochairman of the board of directors. The company sponsored Gilbert's research at Harvard on the bacterial production of insulin. For a time he tried to be both a university professor and an office-bearer at Biogen. But the Harvard authorities could not tolerate this for long, and he finally resigned to become a full-time industrialist (Stevenson, Byerly, 1995, pp. 148-150).

Another example of the significance of commercialization is the discussion in the US scientific community raised by the patent applications made by J. Craig Venter in 1991-1992. At that time Venter was working at the National Institute of Neurological Disorders and Strokes at the NIH with sequencing complementary DNAs (cDNAs) from the human brain. Venter's idea was to focus exclusively on genes and ignore 'junk DNA'. Only genes, not junk DNA, code for proteins by creating messenger RNA (mRNA), and Venter set out cloning copies of such fragile mRNAs. These stronger and analyzable copies are called cDNAs. A sequence from a cDNA clone that corresponds to mRNA is called an 'expressed sequence tag' (EST). By sequencing cDNAs he expected to gain knowledge of actual genes. In addition, he also made sequencing much less expensive.

In its preliminary decision, the US Patent and Trademark Office denied Venter's patent applications. However, Bernadine Healy, Director of the NIH, pressed forward. She was prepared to take a fight for patenting. This resulted in a controversy with James Watson, Director of the National Center for Human Genome Research (NCHGR). Watson was strongly against this rush to patenting. He also argued that what had been accomplished was highly overvalued. As a result of this controversy Watson resigned in 1993. When Healy left the NIH, the new Director, Harold Varmus, withdrew the patent applications arguing that they were "not in the best interests of the public or science". This made patenting of cDNAs a matter for the private sector only. Venter, after leaving the NIH and starting The Institute for Genomic Research (TIGR), and many others have filed such patent applications. The present policy of the National Human Genome Research Institute (NHGRI, formerly NCHGR) is the same as Watson's view. The mere knowledge of a DNA sequence lacks utility and should therefore not be accepted for patent filing. Consequently NHGRI discourages its funded scientists from filing patent applications of cDNAs. However, as Ted Peters notes, here a certain "humorous irony" enters the picture. When the time was ripe for accelerating the sequencing work, the NIH decided to fund someone with a laboratory with the most advanced technology for this kind of work. The grant went to TIGR and its principal scientist, Craig Venter (Peters, 1995, pp. 126-135)! Venter has now left his position as president of TIGR. But his company Celera Genomics has been sequencing the human genome at a higher pace than any other agent, public or commercial.[15]

The commercialization of genetics raises certain issues regarding the social and moral responsibility of scientists. One issue concerns conflicts of interest. It is obvious from the case of Gilbert that conflicts of interests may occur when a person is connected with both university and industry. But what is a conflict of interest? One way of defining it is to say that it is a conflict between professional interests and personal or financial interests. A conflict of commitment, on the other hand, is a conflict between different interests that are linked to a professional role. Such conflicts are internal, while conflicts of interest are external (Werhane, Doering, 1997). How should conflicts of interest in genomics be handled ethically? One way is, of course, to resign from the academic position. This was the solution for Gilbert. However, this is not the only possible solution. In the US universities commonly encourage scientific collaboration with the industry, provided that the collaboration is made public and explicit guidelines are worked out so that there can be no conflicts of interest. I believe that this is the right approach. If commercialization cannot be avoided in genetics, public agendas are better than hidden ones.

There are also several other important issues raised by the commercialization of genetics. Let us take as an example genetic tests that are developed on the basis of the new genetic knowledge.

One fear that was expressed early was that commercialization might lead to premature introduction of new genetic tests. There is a risk for pressure from biotechnological companies to introduce genetic tests even when data are insufficient.

Another risk is that genetic tests will be introduced even if people may not want to use them. This might lead to pressure from biotechnological companies to use them anyhow. This problem is not new. It is a problem we find everywhere in a market economy. Do we actually need all the things that can be bought for money?

In both cases, it should be a part of the moral responsibility of geneticists to counteract the pressure from biotechnological companies. But, of course, there may exist a conflict of interest when they at the same time have shares in the companies.

Commercialization of genetics is also relevant to the question of eugenics. Since biotechnological companies may have interests in reproduction, their products and sales activities may affect what kinds of children are born. This means that such companies may in effect promote a kind of eugenics (see p. 226).

---

[15] See URL: www.tigr.org, and URL: www.celera.com.

Finally, we have the problem of fair distribution of benefits of genome research. If geneticists make profits of their research, does justice not demand that the patients from whom the DNA samples come should also get their share?

Moral Imagination: Gene Patents

Let us now take a closer look at the important issue of gene patenting.[16] As stated above, from the perspective of the geneticists, this is probably one of the most important issues related to social responsibility and research ethics. It is vital that geneticists use their moral imagination in considering this issue from a variety of perspectives.

It is obvious that if DNA sequences could be used as a basis for genetic test kits, genetically tailor-made drugs, and gene therapy technology, then patents on these sequences could be of immense economic value. Two important ethical questions arise in this context:

(1) Should genetic knowledge be allowed to be protected by patents?

(2) If so, exactly what genetic knowledge should one be allowed to patent?

As part of taking responsibility, geneticists would do well to use their moral imagination and reflect on a variety of answers to these questions.

Let us start with an overview of some present regulations and policies.

Regulations and Policies

The term 'intellectual property' refers to a variety of rights associated with inventions, product design, writings, etc. In biotechnology the most important intellectual property rights are patents and trade secrets. Patents give exclusive right to commercially make, use, or sell, an invention for a limited period of time (17 or 20 years, depending on which country we are talking about). In most countries, patent criteria for inventions include that the invention should be novel, useful, and nonobvious to persons with average skills in the field. Ideas, theories, and mathematical algorithms cannot be patented, but processes, such as production methods and diagnostic methods, and products (compositions of matter), such as transgenic microorganisms, transgenic plants, transgenic animals, cell lines, and DNA sequences, can (cf. Eisenberg, 1996).

Let me give some examples of patents in modern biotechnology. In 1980 a genetically modified strain of bacteria was declared patentable by the US Supreme Court. The transgenic bacteria had the ability to break down hydrocarbons and were anticipated to be used in cleaning up oils spills. Because of the insertion of foreign genes it was clear that the bacteria were not naturally occurring and therefore the

---

[16] See Eisenberg, 1996; Svatos, 1996.

Supreme Court found the strain of bacteria patentable. There are also patents on transgenic plants. Examples are the genetically modified tomato 'Flavr Savr' (1994) produced by Calgene, and genetically engineered herbicide-resistant crops. Patents have also been granted for transgenic animals. One well-known example is the 'Harvard Oncomouse' (1988). This mouse has been genetically engineered for increased susceptibility to cancer. Other examples are the sheep Tracy (1990), which produces large amounts of the human protein AAT (alfa1-antitrypsin) in her milk, and the pig Astrid (1992) which was genetically engineered for human transplant. Even human cell lines have been patented. Perhaps the most discussed case is the cell line called 'MO'. In 1976 the leukemia patient James Moore had his spleen removed, the cell line from which was found to be useful for producing valuable proteins. Moore demanded that the cells be returned to him but the California Supreme Court ruled that he had no rights to them once they had been removed from his body.[17]

In relation to genes, there are three conceivable categories for patent protection.
(1) The first category consists of 'expressed sequences tags' (ESTs). As mentioned above, an EST is part of a sequence from a complementary DNA (cDNA) clone that corresponds to messenger RNA (mRNA).
(2) The second category includes methods, for example diagnostic methods, and compositions of matter resulting from identification of disease genes such as BRCA1 and BRCA 2.
(3) The third category consists of individual DNA sequences related to recombinant pharmaceutical products such as erythropoietin, insulin, and human growth hormone (Murashige, 1997).

HUGO accepts patents related to categories (2) and (3). It supports the right of those who have determined the functions of genes to patent their work. However, HUGO is negative to category (1). HUGO finds it unfortunate if the patent system rewards the routine and discourages the innovative. This could be the consequence if patents are granted to those who perform massive sequencing work, while offering more limited rights or no rights whatsoever to those who make discoveries of the functions of the genes, a work that is much more difficult and significant. An early release of sequence data is also important in order to speed up function analysis.[18]

This view is shared by the NIH. The present policy of the National Human Genome Research Institute is that the mere knowledge of a DNA sequence lacks utility and should therefore not be accepted for patent filing. Consequently, scientists are funded on the condition that they refrain from making patent applications of raw DNA sequence.[19] On the other hand, they are encouraged, yes, even obliged, to file gene patents that fulfil the criteria.

---

[17] See Reiss, Straughan, 1996, pp. 133, 166-173; Harris, 1998, pp. 267-268; Erin, 1994.
[18] See URL: www.gene.ucl.ac.uk/hugo/patent.htm.
[19] See URL: www.nhgri.nih.gov/Grant_info/Funding/Statements/patenting.html.

Recently, the same view has also been declared by the EU. In a Directive from 1998, it is maintained that (Directive, 1998, Article 5):

> 1. The human body, at various stages of its formation and development, and the simple discovery of one of its elements, including the sequence or partial sequence of a gene, cannot constitute patentable inventions.

But it is also stated:

> 2. An element isolated from the human body or otherwise produced by means of a technical process, including the sequence or partial sequence of a gene, may constitute a patentable invention, even if the structure of that element is identical to that of a natural element.

In conclusion, there seems to be an emerging consensus on gene patenting. As long as an invention related to a gene is novel, useful, and nonobvious, it may be patentable, and this holds true even if it is a composition of matter produced by technical means, the structure of which is identical with a human DNA sequence.

However, many ethical questions arise. At a very general level, we find the question of whether it is ethically acceptable in principle to patent genes. Moreover, if we accept such patents in principle, more pragmatic problems arise such as, for example: Are patents the best way of stimulating technological innovation related to genes? How could what is rightly common property be preserved while providing possibilities for private development of research results? Let us now take a closer look at these and other related questions.

For this discussion, I stipulate that the term 'gene patents' means patents on inventions related to DNA sequences that meet the criteria of being novel, useful, and nonobvious. I shall start by briefly presenting the main arguments in favor of gene patenting.

Why Patent Genes?

One important ethical principle appealed to in the *pro* arguments is, of course, the principle of beneficence. The key idea is 'no patents, no cure'. The product is costly to make initially but cheap and easy to copy once someone else has made them. Without patents important discoveries of genes will not lead to practical products. Without patents there will be greater secrecy and fewer benefits to health care. Without patents there would be a diminution of novel treatments of poorly treated or untreated diseases. Gene patenting stimulates technological invention and encourages people to give information about their inventions.

Another important ethical principle that can be used in favor of gene patenting is the principle of justice. Gene patents give those who have invested a great deal of money and effort fair compensation. They are rewarded for millions of dollars and many years of work. This is essential because no agency can compensate for the loss of an exclusive market.

Finally, there is a rights-based argument. People who have developed methods or products related to genes have a moral or natural right to them. As soon as someone has made something new out of the original natural raw material, he or she has a right to it. Patents protect these rights. Moreover, these rights are based on the principle of integrity. The methods or products belong to their developers. This linkage should not be broken, and patents protect it.

Specific Objections to Gene Patenting

The objections to gene patenting are of two kinds. The first is specifically directed against the patenting of genes; the second is directed against patenting in general, including gene patenting. I will start with some objections belonging to the first category.

I discern four kinds of specific arguments against patenting of genes. I call them the argument from the sanctity of life, the argument from life as a creation of God, the argument from a common human heritage, and the argument from a commodification of life. All can be interpreted as starting from the principle of integrity and applying it to genes.

Sometimes the arguments are directed against ownership of genes, for instance, expressed in the words "You cannot own life". In these cases, however, they are based on a misconception of what patents are all about. A patent is not a matter of owning. When you own something it can be owned forever, however patents hold only for a limited time. It is not a matter of ownership of a gene. A patent gives an exclusive right to prevent others from commercially making, using, or selling a DNA sequence in a specific application without permission for a certain period.

The argument from sanctity of life criticizes gene patenting for promoting a materialistic conception of life. One version, proposed by Jeremy Rifkin, starts from the assumption that nature is sacred and maintains that natural processes consequently should not be patentable (Rifkin, 1983, p. 252). Other arguments have a more traditional religious character. Some versions stress that products of nature such as genes are creations of God, not of humans. Therefore, they cannot be patented. In 1995, a group of religious leaders representing religions such as Hinduism, Buddhism, Islam, Judaism and different Christian denominations, called for a ban on patenting human genes and genetically engineered animals. They argued: [20]

> We, the undersigned religious leaders, oppose the patenting of human and animal life forms. We are disturbed by the U.S. Patent Office's recent decision to patent human body parts and several genetically engineered animals. We believe that humans and animals are creations of God, not humans, and as such should not be patented as human inventions.

---

[20] Quoted in Peters, 1997, p. 117.

Another argument stresses that genes are a common human heritage like sea beds and famous pieces of art. No one has the right to get a hold on them by patenting. Finally, we have the argument from commodification. It stresses that genes are products of nature, not human artifacts. Moreover, the human genome has an integrity or dignity that should be respected. Consequently, parts of the genome, i.e. genes, are not commodities that can be sold. Gene patenting signifies an unacceptable commodification and commercialization.

General Objections to Patenting Including Gene Patenting

In the debate on gene patenting, there have also been put forward objections to patenting in general that by implication are directed against gene patenting as well.[21] According to one argument, patents promote secrecy and reduce the exchange of scientific information among geneticists. The traditional view in the academic setting is that scientific findings should be shared openly. Patenting conflicts with this view. Patenting is not beneficial to the scientific community.

Other arguments against patenting deny that patents are beneficial to the general public. According to one argument, patents exploit taxpayer-funded research. Private patents are not acceptable when the scientific work is paid for by public funds. Another argument acknowledges that patents allow faster development of some drugs but stresses that patents may also suppress competition and prevent other institutions from conducting beneficial research. Patents have the consequence that efforts are put where money is to be made, rather than where scientific work is most needed. Research on rare and nonprofitable genetic diseases will receive low priority. Moreover, upstream patents may inhibit downstream applications.

Some arguments start from the principle of justice. According to one, patents make important products more expensive and therefore less accessible. The patent implies that the company can sell the product at a higher price with the effect that only rich people can afford it. This is unfair and discriminative to ordinary people. Another argument maintains that it is unfair to put restrictions upon those who wish to use the product or process. A third argument questions whether it is fair that one person, the geneticist, gets all the profit when many patients have contributed with their DNA to make the research possible. These groups of patients should have their fair share of the benefits.

Finally, there are arguments against patenting that focus on certain cultural problems. Concepts of individual innovation and ownership are foreign to many cultures that stress the sharing of community resources. Patenting does not respect the views of these ethnic groups. In other words, it does not respect their integrity. Therefore, fundamentally, this argument refers to the principle of integrity.

---

[21] For a general discussion of patenting, see Merges, 1996, and Pejovich, 1996.

Dialogue on the Results in the Short Term and the Long Term

In the face of this public and international debate on gene patents, what should responsible geneticists do? First of all, they should scrutinize the arguments. In the discussion about gene patenting, we see clearly that commonly accepted principles are interpreted in different ways and that they conflict with each other. This is exactly what could be expected from the point of view of imaginative casuistry. The arguments in favor of gene patenting refer to the principle of beneficence, the principle of justice and the principle of integrity. The arguments against gene patenting also refer to the principle of beneficence, the principle of justice, and the principle of integrity. From this it follows that simple references to general ethical principles do not solve the problem of gene patenting.

As a contribution to the dialogue, I therefore suggest as an alternative approach that responsible geneticists consider certain contextual factors. From the perspective of imaginative casuistry, an appropriate starting point for a consideration of gene patenting is a recognition of the kind of society in which genetics research is carried out. The following contextual features are apparent:

(1) pluralism,
(2) commercialism, and
(3) globalization.

Within the framework of these features of our society, geneticists can realistically reflect on the issue of gene patenting.

First, we do not live in a monolithic society but in an increasingly pluralist one. There are, as a matter of fact, many different and incompatible ethical views in society. There is some general consensus, but only at a very abstract level. There is agreement about ethical principles but when it comes to the interpretation, application, and balancing of these principles, there is often no consensus. The plurality of views is particularly obvious regarding gene technology and gene patenting. The lack of consensus is a fact that must be taken seriously. Given this fact, it is obvious that a reasonable policy on gene patenting cannot give priority to arguments that presuppose certain specific life views that are not self-evident in our pluralistic society. This means that arguments from integrity that focus on the sanctity of life, life as a creation of God, genes as a common human heritage, and gene patents as a commodification of life, should not be given weight to such an extent that gene patents are ruled out in principle. Whether gene patents should be accepted becomes a practical question.

The second characteristic of our society is commercialism. No realistic policy on gene patenting can ignore this fact. Whether we want it or not, commercialism will be with us into the foreseeable future. Our view on gene patenting has to adapt to this fact. As long as there are profits to be made on research, there will be a pressure to seek patents. As the examples presented above indicate, this is true also of genetics research. However, there is a well-established tradition that patents should be approved only on the conditions that the method or product is novel, useful and nonobvious. There is no reason to modify these criteria. So, the question is what the application of these established criteria to genes would imply.

To be sure, commercialism has certain disadvantages. One example is that commercialism to some extent will have the consequence that efforts are put where money is to be made, rather than where scientific work is most needed or where the health needs are the greatest or the most urgent. Another example is that some products may be sold at a higher price with the effect that only rich people can afford them. A third example is that upstream patents may inhibit downstream development. These disadvantages have to be accepted to some extent, although they may also to some extent be counteracted by the government. In the first case, the government may try to direct research efforts in a specific direction by political means. In the second case, the government may provide financial means to help poor people to get the medical products. In the third case, the government may rule by force that knowledge or products with patent protection be made available.

However, despite these disadvantages, it must be recognized that patenting often does have good effects. With regard to health care, effort and money are allocated to research and development that provides better genetic testing technologies and therapies. Thereby, much human suffering may be relieved.

Two other issues raised by commercialization and gene patenting must also be addressed. They concern 'owing', one of the two important metaphors constituting the concept of responsibility. The first concerns the patients who have contributed with their DNA to make the research possible. Do geneticists who make profits due to research on this DNA owe anything to these patients? It is not possible to give individual patients their fair share but it seems quite possible to give it to the group of patients. This can be made as a contribution to their health care.

The second issue is whether private patents are acceptable when the scientific work is paid for by taxes. Should patents on results of research paid for by taxes be allowed or do the geneticists owe it to society not to patent? This is a rather difficult question. It could be argued that it nevertheless is the geneticists, not the general public, who have carried out the research and therefore should be allowed to seek patents. If we accept this argument, the question remains who more precisely should be allowed to seek patents. One possibility is that the universities themselves seek patents, as is the case, e.g., in the US. On the other hand, it is possible to argue that since individual researchers have put a great deal of personal effort on their research, it is also fair that they get the right to seek patents.

Finally, it is obvious that we live in a more and more internationalized world. Globalization is an important feature. Business is an international enterprise, so is genetics research and the development of gene technology. This has to be fully recognized by those who are discussing policies for gene patenting. The argument against patenting that focus on certain cultural problems carries some conviction. It is true that the concepts of individual innovation and ownership are foreign to cultures that stress the sharing of community resources. This should be taken seriously. Before any gene hunting is carried out in these ethnic groups, their 'community consent' must be obtained. If certain genes are identified that can be used for the development of pharmaceuticals, then they should get their fair share of the profit.

My conclusion is that the arguments against gene patenting are not strong enough, and that contextual reasons speak for an acceptance of gene patents. In accordance with the academic tradition, complete openness regarding the results of genetics research would ideally be preferable. But we live in a commercialized world, and if we want to be realistic and take the social context seriously, gene patents can hardly be prohibited. But it is important here to make a distinction between short term and long term goals. Responsible geneticists should reasonably have the long term goal of open access to research findings. They owe this to society, as well as to other scientists. As a long term goal, there should be an open dialogue with respect to the results within the scientific community. In the short term, however, gene patents should be allowed. They hold only for a limited period of time.

How should this view be integrated with scientific practice? First of all, geneticists should not be ashamed of seeking patents. It is acceptable from the point of view of moral responsibility. However, should geneticists also try to affect national patent offices to accept this view? Yes, even that would be a responsible thing to do. If the genetics community in a particular country agrees with this view, which is also the view of HUGO, they should try to find ways of influencing the national patent office to accept it.

However, the question remains to be answered: If gene patents are accepted, how should the problems with other scientists' access to genetic findings be handled? Certain things should be noted here. First, contrary to what many people believe, patents make openness in a commercialized world possible. Without gene patents biotechnological companies would be even more secret with their results. Second, the delay of publication due to patenting is typically only a few months. This means that even if gene patents are accepted, this does not preclude that the whole sequence of the human genome is made generally available rather quickly. The private company Celera Genomics, for instance, has recently released a large amount of information after filing patents for a number of particularly interesting genes.[22] Third, experimental use in basic research of a gene with patent protection does not constitute patent infringement, only commercial use does. Fourth, in the US and Canada there is a 'time of grace' which means that for a certain time after publication it remains possible to seek a patent. The person who has first published his result has the right to get a patent before other people. Such a 'time of grace'

[22] See URL: www.celera.com

could perhaps be instituted also in other countries. It could be a good way of increasing openness while preserving the right to patent. But it is probably not realistic to look for such a radical change of the established patenting traditions in other countries. An alternative could be provisional patents.

In sum, completely open access is not necessarily a short term goal. We may accept gene patents to some extent. But complete openness is clearly a long term goal. This is required if the geneticists want to keep to the prototypical academic tradition. They owe this to society which has given them the opportunity to do the research. It is part of their responsibility for an open dialogue.

CHAPTER FOUR

GENE HUNTING, GENETIC TESTING, AND GENETICALLY TAILOR-MADE DRUGS

1. DIALOGUE IN GENETIC RESEARCH AND APPLICATION

It is a rather uncontroversial view that geneticists should take responsibility for the direct consequences of their research, i.e. those consequences that appear in the research process itself. For instance, they are responsible for the treatment of human subjects participating in the research. More controversial is the view that geneticists should also take responsibility for the indirect consequences of research, i.e. the consequences of publication. I have argued that they should try to affect applications of their findings made by other people.

With regard to genetic research as well as application, dialogue is of key importance. In genetic research, there must be a dialogue between scientists and human subjects on the nature of the research, risks and benefits. In application, there must be a dialogue between scientists and the general public, politicians, and industrialists, on which uses of the results are ethically acceptable and which are not.

I shall now investigate what dialogue and other aspects of moral responsibility could mean more precisely in different kinds of human genetics research and application. This investigation will continue in the next chapter.

'Responding' and 'Owing'

Mapping and sequencing DNA samples from 'immortalized' cell lines that have been completely separated from their human sources and where all identifiers are irretrievably removed do not pose any serious ethical problems with respect to active participation of human subjects. To be sure, when the DNA is first obtained this must be done with the informed consent of the individuals who participate. But the creation of genetic marker maps, physical maps, and determining DNA sequences is pure laboratory work. No human subjects are directly involved. However, the maps and sequences are important tools for further research aiming at the identification of disease genes and function analysis, and in this latter kind of research active participation from human subjects is indispensable.

In all research involving human subjects, certain ethical problems recur. They are related to the two metaphors constituting the concept of responsibility, namely 'responding' and 'owing'. 'Responding' indicates the need for dialogue between geneticists and research subjects. 'Owing' can be interpreted in terms of fair compensation to the subjects for participation in the research. The ethical problems can also be described as related to commonly accepted ethical principles such as respect for autonomy, respect for integrity, and justice. Obtaining informed consent from human subjects is essential, their privacy should be respected, and the medical information should be confidential. Moreover, justice means that human subjects should be treated fairly. However, what does all this mean in human genetics research? Does human genetics research raise any special problems? I shall argue below that there do arise special problems and that the problems get their special character depending on exactly what kind of genetic research is at hand. As examples I choose genetic linkage studies on families and ethnic groups, and research on genetically tailor-made drugs. In relation to each kind of research I shall focus more specifically on problems of dialogue and to some extent fair compensation, and only briefly indicate other problems raised by the research.

Prototypical and Nonprototypical Cases: Autonomy and Informed Consent

Responsible genetics implies dialogue. A key aspect of dialogue in research is obtaining informed consent. Within the scientific community, there is agreement that as a general principle the consent of human subjects must be obtained before doing research. Let us take a look at the concept of informed consent as applied to research on human subjects in general before we turn to genetics research.

The prototypical case of informed consent in research focuses on the individual research subject and includes four main elements, namely disclosure, understanding, voluntariness, and authorization. In addition, there is a precondition, namely capacity.

(1) *Disclosure*. The giving of information is the first important aspect. In general, information should be given to the individual research subject about the aim of the study, methods to be used, anticipated benefits and risks for the subject, the right to withdraw from the study, and criteria for selection of subjects. Perhaps the most important thing for the researcher is to create an atmosphere where the research subject feels at ease to ask further questions. In this way the specific informational needs of each individual can be met.

(2) *Understanding*. This aspect has been stressed increasingly in recent years. It is generally recognized that it is not enough that the researcher thinks that he or she has given correct and sufficient information. The key question is whether the subject has understood. This can to some extent be checked by putting questions to the potential participant.

(3) *Voluntariness.* There are many different ways to violate this aspect of informed consent, direct coercion being the most flagrant. But there are, of course, more subtle violations such as different kinds and degrees of persuasion and manipulation. A threat to voluntariness that clinical researchers should be particularly aware of is when they recruit patients for research. In such cases the patients might feel obliged to participate due to a sense of dependence on the physician or a sense of debt or gratitude for the care and treatment given. It can be discussed what should count as coercion, i.e. whether persuasion, subtle forms of manipulation, or dependency should be included. From the perspective of cognitive semantics, it is obvious that the concept of coercion has prototype structure and can be extended more or less far.
(4) *Authorization.* There might be differences in the quality of the authorization depending on the quality of the disclosure, understanding and voluntariness.
(5) *Precondition: capacity.* A certain minimum ability to understand and decide for oneself is necessary for the principle of informed consent to be applicable. It is a classical issue of controversy in biomedical ethics whether and under what conditions it is justified to involve children and mentally incapacitated persons in research.

The principle of informed consent can be justified in different ways. The two most common justifications are related to the principle of nonmaleficence and the principle of respect for autonomy, respectively. In the early years of research on human subjects informed consent was commonly justified with reference to the former. Consent was seen as a way of minimizing potential harm. In recent years, however, justification with reference to the principle of respect for autonomy has become the dominating one. Autonomy would be without substance if human subjects do not receive information enough to be able to decide for themselves whether to participate.

Recently, however, the individualistic notion of informed consent has been criticized with reference to genetics. This suggests, in terms of imaginative casuistry, that in genetics research we face nonprototypical cases with regard to informed consent. It is possible that the proper unit of consent is the family, not the individual. It is the autonomy of the family that should be respected. This is so because information about one family member's genes implies information about other family members' genes as well. In the first chapter I showed the metaphorical character of the concept of autonomy. The term was originally used about the self-legislative power of the ancient Greek city-states. This political concept has long been used metaphorically about the self-determination of individuals. Now, we see an extension of the metaphor from individuals to families. This is an interesting conceptual development.

Moreover, the metaphor may be extended even further. In the discussion about genetic studies of different ethnic groups around the world, it has been suggested that 'community consent' should be obtained before the research starts. This community consent presupposes that we can talk about a kind of community autonomy. This means that we come close to the original meaning of the term, namely the autonomy of the ancient city-states.

Furthermore, autonomy has also been extended to future generations by some critics of germline gene therapy. An argument against germline gene therapy is that it is unacceptable because the consent of future people has not been obtained before the modification is carried out. 'Future generation consent' presupposes 'future generation autonomy'.

In some of the sections below, I shall discuss these new suggestions about informed consent and autonomy. The issue of 'future generation consent', however, I shall leave to the next chapter. But I shall also deal with some other ethical problems raised by genetics research involving human subjects.

2. GENE HUNTING IN FAMILIES

Genetic linkage studies raise special ethical problems largely unanticipated in other research involving human subjects. Most of the problems are related to dialogue in the wide sense: recruitment and consent, withdrawal, consent for new studies, disclosure of results, and publication. It is essential that responsible geneticists use their moral imagination in considering these problems, empathizing with the persons involved and envisioning different alternatives of action. For the most part, I shall stay content with presenting the problems, and refrain from giving any suggestions about their handling. The idea of family consent, however, will be discussed rather extensively, because it raises a key problem with regard to dialogue in this kind of research.

In most genetic family studies the research subjects are recruited from clinical practice settings in which family members have been involved in genetic counseling, and where the family pedigree is at least partially established. As the study progresses more members of the extended family are recruited. At a practical level, this means that blood samples for DNA analysis are collected from all these people.[1]

Recruitment and Consent

A key problem in gene hunting targeting families concerns obtaining consent from family members. Should the family members be recruited individually or should the family be recruited as a whole leaving the family's internal dynamics intact? The traditional view on recruitment and consent has focused on individual autonomy. But perhaps this model makes us pose the wrong questions?

A closely related problem concerns who should recruit other family members as research subjects. Should it be the researcher or the family member already involved? This is a delicate matter. It is one thing in a purely clinical context to

---

[1] Many of the problems analyzed below are discussed in Frankel, Teich (eds.), 1993, and Juengst, 1994.

encourage the individual to contact other family members at genetic risk. This is difficult enough. But it is even more difficult to recruit members to a research study that aims at increased knowledge rather than clinical intervention. It is also obvious that the pressure from other family members to participate constitute a threat to voluntariness.

What about the content of the information given to members of the extended family? Is it acceptable to disclose information to recruited members of the extended family about the disease or disorder under study? If the study concerns a socially sensitive disease or disorder, for example, a psychiatric disorder, this might lead to stigmatization of certain members of the family. Moreover, there is the problem concerning whether to disclose psychosocial risks connected with participation in the study. The medical data obtained through the study might be requested, for example, in insurance. This might lead to discrimination and stigmatization of the research subjects. Participators may also have problems with understanding the psychosocial risks.

All this means that in genetic family studies even the element of authorization becomes problematic. Should it be a matter of individual authorization or should the family as a whole give its authorization? Moreover, is it acceptable to involve children and the mentally incapacitated? Can they be involved without consent? Since they do not have capacity for consent, this means that other ethical principles are relevant. Should a kind of vicarious family consent be obtained?

Withdrawal

Other problems are related to withdrawal from research studies. It is a commonly recognized right of the research subjects to be able to withdraw from the study if he or she wants to. Consent does not mean that it is not possible to change your mind. However, does this right to withdraw from the study oblige the scientist to remove information from the pedigree and discard their DNA stored in the freezer? Moreover, this raises the more general question about who owns DNA samples and genetic information.

The traditional view is that DNA samples, like other donated human biological material, is material over which the research subjects no longer hold a claim. An important objection to this view, however, is that unlike most donated tissue DNA samples in a family collection give them scientific value precisely to the extent that they continue to represent their donors. Without identification they are scientifically worthless. One solution for the scientists could be to discard the DNA samples but retain control over the anomymized genetic information. A problem with this view is that even anonymized information can be recognized.

Consent for New Studies Using Stored DNA Samples?

One further problem is raised by new studies using stored DNA samples.[2] Is such studies acceptable without new consent? In this context, it is important to make a distinction between studies where all identifiers are irretrievably removed and studies where the identifiers are not removed. In the former case, it might seem self-evident that no new consent is needed. Traditionally, research on unidentifiable human tissue has been exempt from consent procedures. Most cell lines are distributed within the scientific community without personal identifiers. However, this view is increasingly coming under pressure. It is almost impossible to 'anonymize' the full family collection and at the same time use it for gene hunting. The reason is that DNA samples are related to each other in family studies and can be used to identify a particular family. Moreover, even for studies that do not need knowledge of family relationships, it may in the future be possible to reidentify individuals by comparing the research collection with their DNA fingerprints (Juengst, 1994.).

But there are also other reasons why consent should be obtained. Even if there is no direct harm in terms of psychosocial risks for the research subjects, there might be indirect harm. The research subjects might want to limit the use to noncommercial use. They might want to forbid use for research on certain disorders, particularly if it is stigmatizing for a specific group. One example could be research for a gene for alcoholism. Moreover, the research might violate cultural or religious beliefs.

What about new studies in which identifiers are not removed? Is new consent necessary in such studies? According to the traditional view, there is no need for consent if the research subjects in their original consent have given blanket consent in the sense of admitting that the samples can be used "for further research". A problem with this view is that other family members are identifiable. With this in mind, an increasingly common view is that new consent should be obtained in the same way as in the first case (Juengst, 1994). This means that blanket consent is unacceptable.

A common objection to the demand for obtaining consent in new studies using stored DNA samples is that it takes too much time and effort. The scientific studies are too important and the issue of consent too unproblematic. An possible answer to this objection is based on the observation that contacting people always entails some costs, and stresses that there must be proof that obtaining consent would be so burdensome and expensive that the research could not go forward if it was necessary (Clayton et al., 1995).

Disclosure of Results to Participants?

---

[2] See, for instance, Clayton et al., 1995.

There are also ethical problems that are related to the disclosure of research results. Most genetic family studies begin as basic research projects. Do geneticists have a duty to inform research subjects about the results of the studies? Is there in this respect a difference between preliminary and confirmed results? Are geneticists obliged to provide clinical follow-up services including genetic counseling?

It is obvious that no information about preliminary results should be given to research subjects. There is clearly a need to confirm research findings, e.g. a gene–disease association, before disclosing the results. It could also be argued that a researcher looking for a gene–disease association for the first time would do well to inform subjects that they will not receive the results. However, once the results have been confirmed the researcher has to decide whether to recontact the subjects. Information whether the confirmed results will be disclosed should be included in the original informed consent disclosure. When deciding whether to disclose confirmed results, relevant aspects to consider are the severity of the genetic disease, the probability of phenotypic expression, and the availability of treatment.

If it is decided that the confirmed results will be disclosed, it is clear that the researcher has a responsibility to provide genetic counseling. In practice there are two options. One is based on a separation of the scientific role and the clinical role. In this case, the geneticists refer the research subjects to a genetic counselor. Another alternative is based on an integration of the scientific and the clinical roles. The researcher himself or herself provides the counseling.

Publication of Pedigree Studies

Another important problem concerning disclosure of scientific results concerns publication of pedigree studies. The general issue is how to protect the privacy of the research subjects. There are many specific problems in this context. Even if the pedigrees are anonymized the identities may be recognized by family members and those who know them. This means that carrier status of other family members, adoptions, stillbirths, and misidentified paternity can be revealed.

The traditional way of handling this problem is to disguise the identities in the published pedigrees by changing the gender and birth order, and keep the original data available only to other scientists who are trusted (Juengst, 1994). A more radical policy has been suggested for cases of very uncommon diseases with few research subjects, namely to publish results without the pedigree. However, in such very rare cases, the rarity itself may make identification possible.

Justice: Fair Compensation

Other ethical problems raised by genetic family studies concern fair compensation. Enormous profits might be made once a disease gene has been identified. Do the geneticists owe the participants any compensation? Yes, it seems fair that patients who have contributed to the research should be compensated in one way or another. However, it does not seem reasonable to let the family members share the commercial profits. To be sure, they have only contributed with blood samples. They have not carried out any research. One alternative could be to compensate the family members through financial support of their treatment and care.

Dialogue: The General Problem of Family Consent in Genetics

Robert Wachbroit has criticized the traditional understanding of confidentiality as applied to genetics. He suggests a new framework for thinking about confidentiality where the family, not the individual proband, is considered to be the patient proper (Wachbroit, 1993). As far as I can see, his view has implications not only for the view on confidentiality but also for the view on consent. The setting of Wachbroit's alternative framework is clinical practice. But it seems to have implications for the research setting as well.

What generates the ethical problem is the fact that genetic information may include information not only about the individual from whom it is obtained (the proband) but also about other family members. An example discussed by Wachbroit is Duchenne's muscular dystrophy, which is a recessive, x-linked disease. If genetic information about such a disease is not only about the individual, to whom is confidentiality properly owed?

According to the traditional view, health professionals owe confidentiality to the individual proband only. They are neither permitted nor required to disclose any genetic information to other family members without the consent of the proband. This individualistic framework is criticized by Wachbroit. He suggests that because genetic information might be not only about the individual but about the entire family, the family should be considered the 'patient' proper. From the perspective of cognitive semantics, this can be described as a metaphorical extension of the concept of a patient from the individual to the family. Within Wachbroit's alternative framework, the issue of confidentiality changes radically. The proper subject of confidentiality is not the individual but the family. The problem of the confidentiality of the individual suddenly vanishes. Health professionals do not owe individuals any confidentiality. They only have a duty to protect the confidentiality of the family as a whole. This means that they are not only permitted but also required to disclose the genetic information to the other family members. In terms of cognitive semantics and imaginative casuistry, Wachbroit extends the concept of consent from the individual to the entire family. He can also be said to extend the concept of autonomy from the individual to the family, because the reason for obtaining consent and protecting the confidentiality of the family is ultimately that the family is the proper unit of autonomy. It is the autonomy of the family that should be respected, not the autonomy of the individual.

In order to understand Wachbroit, it is important to make a distinction between a public health approach and an individualistic, client-centered approach. The latter presupposes individual autonomy and individual consent from the start. The former approach, however, better fits the ideas of family autonomy and family consent.

There are several problems with Wachbroit's proposal. Wachbroit himself discusses three problems. The first concerns the unit of the 'patient': when the patient consists of a number of individuals, who speaks for the patient? Family members often disagree. The second problem concerns where to draw the line regarding whom the afflicted relatives are who constitute the patient. The third problem concerns the force of the obligations of the health professionals. They are not merely permitted to disclose genetic information to the entire family, but are required to do so. But people have a right not to know. While acknowledging that each of these problems raises concerns about the implementation of the alternative framework, Wachbroit maintains that they do not constitute decisive arguments against it. They are 'hard cases' that we have to deal with, but they do not invalidate the framework as a whole.

Wachbroit does not take these problems seriously enough. They are not just 'hard cases' but problems that indicate that the framework is unacceptable. First of all, Wachbroit underestimates the prevalence of conflict of interest within families. Health professionals should not ignore the internal dynamics of family life. They have to recognize that not all families are devoted to democratic decision-making. Even if all families were, the oppression of minorities within a democratic system is a well-known and serious problem. The interests of individual family members might be suppressed by the dominating ones.

Furthermore, the problem of where to draw a line is more serious than Wachbroit seems to acknowledge. It is difficult to draw a line regarding who the afflicted family members are. The line risks to be arbitrary and this problem should be taken seriously. Why stop with the family? Why not treat the society as a whole as a patient? Genetic disease in certain individuals and families might be viewed as a disease of the 'body of society'. However, if we start viewing the population at large as patient, we come close to eugenics. The gene pool of society becomes the patient proper. Does this mean that we are justified in sacrificing individual confidentiality for the good of the gene pool of society? From a historical point of view, respecting the autonomy of individual patients seems to be a less bad alternative. The history of eugenics is too abdominable.

Finally, the idea that health professionals could be required, not only permitted, to disclose genetic information to other family members is also open to objection. This is an unattractive implication of Wachbroit's alternative framework. He explicitly stresses that the aims of public health and individual health sometimes conflict and argues for the priority of the public health objectives. From this perspective, it simply makes no sense to talk about the confidentiality of the individual. This also comes close to eugenics. For historical reasons, eugenics should not be accepted. A high appreciation of individual autonomy is the best safeguard against and antidote to eugenics as traditionally understood (see pp. 241, 246).

Wachbroit's alternative framework concerns clinical practice, but it seems possible to apply it to research as well. However, my assessment would be the same: it should not be accepted. The family-based nature of the genetic linkage studies of families does not require family-based consent.

Conclusion

Dialogue is central to moral responsibility in genetic family studies. Many problems in this kind of research are related to dialogue in the wide sense. A special problem is raised by the fact that genetic information may include information not only about the individual from whom it is obtained but also about other family members. This may suggest that family consent, not individual consent, is appropriate. This view, however, is rejected. It is vital, though, that responsible geneticists take the possibility of family consent under consideration.

3. GENE HUNTING IN ETHNIC GROUPS

There has been a repeated call to launch a global Human Genome Diversity Project (HGDP) to map the genetic variation within the human species. The HUGO's HGD Executive Committee was formally established in 1991 with Luigi Cavalli-Sforza as chairperson and organized into different regional committees (Zilinskas, 1997, pp. 62-63).[3] However, the Project as a global cooperative effort has not truly begun yet. For instance, no sampling and analysis has been carried out in North America, South America or Oceania. However, Europe has been rather active. Here the project is called "Biological History of European Populations", not HGDP. Moreover, the governments of China, India and Pakistan are funding HGDP research. Cavalli-Sforza concludes that "[t]his makes about one third of the world" (Cavalli-Sforza, 1997, p. 56).

The aim of the HGDP is to collect, preserve, analyze, and make available genetic information from ethnic groups all over the world. Concretely, the project is a matter of collecting blood samples from different ethnic groups, preserving these samples in repositories, doing genetic analyses of them, and making the results of these analyses available through computerized databases. In order to make the genetic analyses possible, the cells are multiplied and in a sense immortalized.

---

[3] See also URL: www.stanford.edu/group/morrinst/hgdp/summary93.html.

However, while the HGDP has got a slow start, there is already another kind of gene hunting going on in various parts of the world. Ethnic groups, many of them in the Third World, are approached in the search for genes for genetic drug research.[4]

These two kinds of gene hunting raise to some extent similar ethical questions. But there are also important differences. One is that while both kinds of research will probably generate enormous commercial profits, this motive is more obvious in the gene hunting for drug development. Below I shall focus primarily on the HGDP, but much of what I say will be applicable to gene hunting for drugs as well.

The ethical questions raised by the HGDP are quite similar to those raised by genetic family studies. Examples are problems of dialogue related to consent, privacy and confidentiality, and problems of 'owing' in terms of fair compensation for participation. However, the HGDP raises a unique combination of such issues. The problems get their special character depending on the particular kind of genetics research done within the HGDP, namely studies of small ethnic groups many of which belong to the Third World. Many live on the edge of extinction and many have had very bad previous experiences of white people in terms of discrimination, oppression and exploitation.

I shall start by discussing the justification of the HGDP. Then I shall focus on some specific ethical problems raised in the process of research and publication. These are all linked to the general question of treatment of human subjects in research.

Justification of the HGDP: Scientific and Moral Imagination

A research project can be justified in terms of scientific value as well as social value. In judging a project, a balancing is needed between these two kinds of consideration. In the previous chapter, the analytical distinction as well as the balancing approach were applied to the Human Genome Project. I shall now do the same with regard to the HGDP. As stressed throughout this book, my suggestions should only be viewed as contributions to a dialogue with geneticists and others. What is vital from the perspective of moral responsibility in science, is that geneticists take all these different arguments and perspectives into consideration using their scientific and moral imagination, and make their own balancing.

Let me start with the expected scientific value. First, the project is expected to provide knowledge about the biological development and history of the human species. If the findings from archaeology, linguistics, history and other disciplines are integrated with the genetic findings, we will get a much richer picture of our past. We will gain knowledge of true biohistorical and phylogenetic relationships.

---

[4] A controversial example in a European country is the company Decode Genetics' research on the population of Iceland. See URL: www.decodegenetics.com.

Another argument that has been put forward is that the project will correct the bias of the HGP. The majority of the genes being sequenced come from a limited number of North Atlantic European-American lineages. By studying the genes from other ethnic groups all over the world this unrepresentativeness will be counteracted. For instance, the differences within the group of Afro-Americans and within the group of Africans in general will be highlighted. We will get a more adequate picture of the genetic similarities and differences between ethnic groups (Jackson, 1997).

The project is also expected to be of great biomedical value. It will provide information that helps us in understanding and treating diseases with genetic components. Genetic testing, genetic drug design, and gene therapy will profit from the new knowledge.

Another argument concerns the issue why the HGDP should be carried out now and not later: the number of ethnic groups are diminishing at a fast pace. If we shall ever get knowledge of many of the smaller ethnic groups, we have to study them now. The project is very urgent. If we wait a few decades, the project will reach much poorer results.

If we now turn to the expected social value of the HGDP, we find that the value most strongly stressed is the health benefits. The biomedical knowledge gained through the project will lead to improved global health. Not just Europeans will benefit but people from many different ethnic groups all over the world.

Another possible social benefit that is sometimes stressed is that the knowledge gained by the HGDP will be of importance in fighting racism. By providing information about true biohistorical and phylogenetic relationships and about similarities and differences in the genetic make-up of ethnic groups, the unfounded views of racists can be counteracted. The differences within ethnic groups are expected to be bigger than the differences between them.

The HGDP has, of course, also met a great deal of criticism. Few people, however, question the scientific value. The objections focus on other aspects. One objection, quite opposite to the view just mentioned, is that the results might be used by racists. In studying the genetic differences between ethnic groups, we might find differences in genetic dispositions for, for example, cognitive abilities. Such knowledge might be used for political purposes of a dubious nature.[5]

Another objection focuses on the manner in which some scientists collect the genetic information, despite the high ethical ambitions of its initiators. 'Bleed and run' collecting is a fact,[6] however, probably more obviously so in gene hunting for drugs than in the HGDP. Informed consent is not obtained from the community. No or very poor information about the point of taking blood samples are given. Often two samples are taken from each individual, one for medical reasons and one for

[5] Cf. the discussion of *The Bell Curve* on pp. 111-114, 227, 234.

[6] Cf. the Channel Four film *Gene Hunters* directed by Ian Taylor (ZEF Productions Ltd., Brighton Media Centre, 9-12 Middle Street, Brighton BN1 1AL, UK).

scientific reasons. The latter is taken without explanation or consent. The former is taken with the expressed goal of helping the individual. But this actually only conceals the real reason, namely to get blood samples for genetics research. This objection deals with the methods used by some scientists, not with the scientific goals as such.

Moreover, many communities have very clearly stated that they do not want to participate in the project. If they do not consent, the research should not be carried out. This refusal must be taken seriously. Many of the communities that have been informed in one way or another have strongly criticized the project.[7]

Finally, if the intention is truly to help small ethnic groups in the Third World, there are many better ways of doing this. Many of the ethnic groups are on the edge of extinction and the argument that this justifies rapid collecting can be turned upside down. Why not help these people to survive instead? Moreover, probably only rarely will the ethnic groups themselves get a fair share of the benefits of future medical applications.

How should these different kinds of justification and criticism be evaluated? I find no problems with the arguments put forward as justification. Scientifically, the genetic knowledge obtained will give a more adequate picture of our past and also correct the bias of other human genome research. In addition, it will increase our biomedical knowledge. All this is, of course, valuable. We have at the moment a unique opportunity to obtain genetic information from ethnic groups all over the world, some of which will soon disappear.

Socially, the project will probably have some positive effects on health care, even if these are difficult to anticipate. One possibility could be that genes for immunity against certain diseases are found. Another is that drugs are developed on the basis of the new genetic knowledge. However, there is a risk that the ethnic groups will not get so much of these medical benefits. This criticism is not enough to support an inhibition of the project but it should be a memento and an incentive to further efforts to give the Third World a more fair share of health care benefits.

Regarding racism, it seems evident that the results may be used as an important weapon in the struggle against it. We have reason to believe that the differences within ethnic groups are bigger than those between them. However, even if the genetic differences between ethnic groups are small, they may be referred to by racists. But this abuse is not reason enough not to support the project. It is an abuse that should be fought by the genetic community. In the previous chapter, I have already discussed the moral responsibility of geneticists to fight such a misrepresentation of genetics.

The argument that there are better ways of helping ethnic groups in the Third World than implementing the HGDP also carries some conviction. However, as we have seen, this kind of research can be pursued for purely scientific reasons.

---

[7] See, for example, URL: www.ipcb.org/resolutions/kunadec.html.

Moreover, you can always argue that the money should be used for other purposes than scientific research. But this does not seem to be a politically realistic option.

Finally, we have the 'bleed and run' objection and the 'refusal' objection. These criticisms must be taken very seriously. It is bad for the credibility of the project if nonserious scientists use obviously unethical methods. One thing to do could be to develop ethical guidelines specifically for the HGDP. Actually, this has already been done. The North American Regional Committee of the HGDP has developed a *Model ethical protocol for collecting DNA samples* which, on the whole, is very good.[8] The problem is that some scientists do not take these guidelines or other more general guidelines for biomedical research seriously. This is totally unacceptable. The question is, however, how 'bleed and run' collecting and neglect of refusal should be prevented. One possibility could be that scientific journals require that the ethical aspects should be presented. A section with an ethical account could even be mandatory for scientific articles presenting research involving human subjects. This might be effective.

In conclusion, the objections to the value of the project are not tenable enough to justify that the project should be stopped. Moreover, the scientific and, to some extent, social value of the project are obvious. There are, however, several ethical problems that need to be handled within the project. We have already met some of them. Below I shall discuss a few other ethical issues. They indicate that certain measures have to be taken in order to make the project ethically acceptable. If these measures are not taken the project should not be carried out.

Before we turn to these problems, let me first briefly comment on the role of the *Model ethical protocol* in the promotion of the HGDP. The existence of this protocol could be interpreted as way of legitimizing the HGDP and avoiding bad reputation. We have already seen that the ELSI programs may have this legitimizing function in relation to the HGP. It is obvious that the protocol may have this function as well. However, in neither case this is something bad. All science is conducted in a social context and from a sociological perspective this legitimizing function cannot be left out of account. What is important, however, is that the initiators openly acknowledge this function and that the normative suggestions are openly discussed. ELSI and the *Model ethical protocol* have a legitimizing function but this is not their only function. They can be used to actually make genetics research more morally responsible.

Dialogue and Autonomy: Individual Consent vs. Community Consent

---

[8] See URL: www.stanford.edu/group/morrinst/hgdp/protocol.html

One ethical problem concerns the meaning and scope of consent in the HGDP. In the section about genetic family studies, we met the problem of individual consent versus family consent. In relation to the HGDP, we have a similar problem of dialogue but with larger dimensions. Most people would agree that consent should be obtained from the individuals from whom blood samples are to be taken. But should the scientists also obtain consent from the ethnic group as a whole? What are the reasons for such 'community consent' and are these reasons acceptable? Is the notion of 'community autonomy' on which the idea of community consent is based acceptable? Moreover, if community consent is necessary, how should it be obtained in practice?

The level at which consent should be obtained may differ depending on which ethnic group scientists intend to study. This means that the question of the level of consent should can and should be answered only by the ethnic group itself. Moreover, before consent can be obtained contact must be established. How this should be made can also be a delicate matter. The ethnic group may require that certain rituals or procedures should be observed. In order to get information about the appropriate level of consent and the appropriate manner of making contact, the researchers in the project have to ask in advance other people who have been working among them, for instance, anthropologists, doctors, or missionaries.

What reasons can be put forward for and against obtaining community consent? The main reasons in favor refer to the principles of autonomy and nonmaleficence. The principle of respect for autonomy is traditionally applied to the individual. The individual is viewed as a self-governing being and should be treated as such. However, in our analysis of consent in genetic family studies, we discussed the idea that the appropriate subject of consent in such studies is the family, because information about one family member's genetic make-up to some extent may imply information about the genetic particularities of the others. The family is seen as a self-governing whole whose consent should be obtained. In a similar way, we now see that some ethicists suggest that the community should be viewed as a self-governing whole and that the consent of this whole should be sought. Note that 'self-governing' here might mean legally regulated self-governing. However, it refers more often to nonlegally regulated self-governing. Perhaps such communities, to the extent that they have a clear organization, can be viewed as nongovernmental organizations (NGOs). Sometimes such NGOs are only faint remainders of precolonial times. The respect for the autonomy of the ethnic groups implies respect for their culture. This means that their ideas about the appropriate level of consent should be respected. The culturally appropriate governing authorities should give their consent. Commonly, community consent should be obtained from the community leaders such as the chief or a council of elders. Sometimes it should be obtained from religious authorities. Moreover, it should be noted that the community may not be as individualistic as Western communities.

The principle of nonmaleficence may also be referred to as a reason for seeking community consent. The ethnic group may have been discriminated against and oppressed in the past. It might even have been subject to genocide. These historical experiences may make the population reluctant to participate. In order to avoid a case in which the group feels more harm, community consent might be necessary to seek. In addition, community consent might be the best protection from exploitation in the future. The community leaders get at least some control over what is happening in their community.

Another reason for obtaining community consent is that the population-based nature of the research in itself requires population-based consent. This is a reason stressed by the *Model ethical protocol*.[9] The project intends to study populations, not individuals. Therefore, the consent of the population must be sought as well, not only the consent of the individuals.

However, there is an important conceptual objection to the idea of community consent. Eric Juengst has stressed the importance of making a distinction between demes and and ethnic groups. 'Demes' are groups of individuals who are more genetically similar to each other than to any other individuals. 'Ethnic groups' are self-identified social groups. The subjects of study of population genomics are demes, not self-identified social groups. Therefore, it is a bad idea to let the self-identified social groups serve as gatekeepers in population genomics research. For conceptual reasons, they cannot function as proxies for demes. Moreover, this would send the wrong message that they are reducible to demes. It is a generally acknowledged fact that the genetic differences between ethnic groups are much smaller than the genetic differences within ethnic groups. The demes to which individuals belong may not coincide with the self-identified social groups. At least, we cannot know this in advance (Juengst, 1998a).

An ethical objection to the idea of community consent is that community consent violates the rights of individuals to decide for themselves whether to participate in research projects. The notion of community consent conflicts with commonly acknowledged human rights. In cases of refusal from the community leaders, it violates the rights of an individual who wants to participate nevertheless. In cases of consent from the community leaders, it violates the rights of an individual who does not want to participate. In this latter case, it is, of course, possible to supplement community consent with individual consent so that the individual has to give his or her consent as well.

In addition, the nesting of local groups within larger social categories implies that no group have moral reach to decide for all. This erodes the idea of community consent.

Let me comment on each one of these arguments starting with the argument from the principle of autonomy. To extend the notion of autonomy to communities is to

[9] See URL: www.stanford.edu/group/morrinst/hgdp/protocol.html

some degree quite acceptable. Actually, it is in accordance with original notion of autonomy in ancient Greek city-states. To the extent that population genomicists seek to use self-identified social groups in their research obtaining community consent is legitimate and vital.

In a similar way, the argument referring to the principle of nonmaleficence is also to some extent acceptable. There are real risks for discrimination of group members. To the extent that community consent reduces these risks, obtaining such consent is essential.

However, to the extent that population genomicists do not use self-identified social groups as research subjects the argument from autonomy and the argument from nonmaleficence become less important. If they design their study of human genetic variation in terms of random, but systematic, sampling in order to identify demes, the need for consent disappears or at least the content of information gets a different character. The reason is that the researchers do not know in advance which individuals from which social groups belong to which demes. In such research there is no good reason to let self-identified social groups play the role of gatekeepers. Moreover, this means that the argument that the population-based nature of the research in itself requires population-based consent is weak. It holds only to the extent that 'populations' are identified with self-identified social groups and these social groups are the actual subjects of study.

The argument from individual rights and the difficulty of finding a morally acceptable group to function as gatekeeper when groups are nested in larger social structures should be taken seriously. Violations to individual rights should not be taken easily even if the individuals belong to nonindividualistic cultures. Similarly, suppression of groups within larger social categories is also a risk that should not be neglected.

In conclusion, to the extent that population genomicists seek to use self-identified social groups in their studies community consent should be obtained, although there are some hard cases related to individual rights and finding morally acceptable 'spokesgroups' when the groups belong to larger social communities. However, to the extent that they do not seek to use such groups, obtaining community consent is neither necessary nor good.

However, the question arises: what should scientists do if they are refused? The simple answer is that they should abstain from carrying out the research project. However, as Juengst points out, what is needed is to find alternatives to using self-identified social groups as subjects of study in population genomics. In this way, the question of community consent and community refusal disappears or at least the information to be given becomes different.

So, for various reasons, the truly important question is how to proceed if self-identified social groups are not to be used in population genomics. What alternatives are there? Juengst stresses that in theory, the research could be carried out by random, but systematic, sampling from blood banks, hospital drains, and battlefields around the world (Juengst, 1998a). However, this may also involve difficult problems with regard to informed consent. In addition, the question remains how to approach groups from which no samples are as yet obtained. I would suggest that if such a group is to be approached, it must be informed that the target for the research is a deme, not the ethnic group as such. Moreover, a possible refusal must be respected.

Other Questions Related to Community Consent

To the extent that community consent is to be obtained, what should be the content of the information disclosed to individuals and community leaders? For obvious reasons there are special practical problems involved in giving information about modern genetics to people with little or no scientific education and perhaps also with no education in the Western sense whatsoever. The *Model ethical protocol* quotes CIOMS's definition of the scope of informed consent which focuses on the information that a 'reasonable person' would consider material to making a decision about whether to consent.[10] I am critical of this formulation. The standard of disclosure cannot be defined independently of the context. The definition must be contextual. What is appropriate information depends, among other things, on the cultural context of the person to consent. This becomes particularly clear regarding the HGDP. What is a reasonable person according to the populations to be studied? What is adequate information to people with a culture that deviates from the Western one? For example, how does one explain that the target for the study is a deme, not the ethnic group as such?

Regarding information about the purpose and nature of the study, it might be necessary to use analogies. These could allude to common human experiences or to the community's own conceptual framework. With respect to information about the risks of the individual's participation, it might be necessary to explain that the blood will not be used in any kind of magic. In both cases, it is vital to cooperate with anthropologists and others who are familiar with the culture.

A problem might emerge if the community leaders force people to participate. In such cases, the scientists should try to use only those who can be expected to be volunteers. They must stress that participation is voluntary and that they have the right to withdraw from participation.

Finally, we have the question of what form consent should have, written or oral. It seems obvious that it need not necessarily be written because some ethnic groups might distrust written documents. Moreover, limited literacy may make documents meaningless.

---

[10] See URL: www.stanford.edu/group/morrinst/hgdp/protocol.html

Integrity: Privacy and Confidentiality

As the issue of consent must be solved on the basis of knowledge about the culture of the ethnic community studied, so must the issue of privacy. The protection of privacy in the collecting process must be based on the specific norms of the population from which samples are collected. In some communities, it would be interpreted as rude to take the blood samples in front of the family or neighbors. In other communities, it would be rude not to do so.[11]

Regarding the confidentiality of the results, we can make a distinction between individual confidentiality and community confidentiality. Both raise ethical problems. In the former case, should complete individual confidentiality be the goal? One could argue that since the subjects of study are populations, not individuals, information about individuals should be completely excluded. However, scientific analyses might need information about individuals in terms of cultural background or in terms of family relationships. The information about individuals should not, however, be available in the databases. Access should require special application and only be given in clearly justified cases.

Complete community confidentiality is contradictory to the goals of the HGDP. Information about which population the DNA samples come from is absolutely essential to this kind of research. On the other hand, information about exactly which subgroup of people within a community from which the samples are taken is seldom necessary; neither is information about the exact location of such a subgroup.

Justice: Fair Compensation

In the discussion of genetic family studies, I argued for the importance of compensating the family members from which samples are taken. I shall now briefly discuss the same issue in relation to communities participating in the HGDP. The provision of benefits can be justified on the basis of the principle of justice. This principle demands that researchers help populations that help them. It is also important to make a distinction between individual compensation and community compensation. But what exactly do the geneticists owe individuals and communities participating?

The provision of benefits is clearly stated in the *Model ethical protocol* and this is very good. The protocol explicitly stresses that the benefits should be appropriate in nature, scale, and distribution.[12] First, it is important to demarcate against bribery. Second, the amount of benefits should be limited. Too much can be equivalent to coercion. The subjects cannot refuse. The benefits would be too tempting. One kind

[11] See URL: www.stanford.edu/group/morrinst/hgdp/protocol.html
[12] See URL: www.stanford.edu/group/morrinst/hgdp/protocol.html

of benefit that is particularly appropriate is medical services. Another kind could be supplies. Third, distribution should be appropriate and this can also be a difficult matter. To give an important gift to a person of low rank can be devastating not only for the social system but also for that individual. Appropriate distribution of benefits requires good knowledge of the culture at hand.

The research within the HGDP may lead to results of commercial value. However, it should be noted that it is not one of the goals of the project; it is only a possibility that cannot be ruled out. I have already discussed issues of commercialism and patenting in relation to the HGP and genetic family studies. The commercial aspects exacerbate the problems of fair compensation. How should the ethnic groups get their fair share of possible profits? The *Model ethical protocol* suggests three possibilities.[13] The first is to require written permission from the community. Another could be to pay a percentage royalty. A third possibility would be to negotiate a payment with a trustee for the community, e.g. UNESCO. Without taking a stand regarding which of these alternatives should be preferred, it is important that in all cases, the compensation should be to the community as a whole, not to individual members.

Conclusion

Many ethical problems in genetic studies of ethnic groups, such as the Human Genome Diversity Project, concern dialogue. Special problems in this respect arise because many third world cultures are nonindividualistic and many of them have historical experience of Western exploitation. This makes it necessary to obtain community consent before obtaining individual consent. It also makes it necessary to find creative ways of giving the community fair compensation for participating in the studies.

4. GENETIC TESTING AND GENETIC COUNSELING

Gene hunting aims primarily at the identification of disease genes. This holds true for family studies as well as studies of ethnic groups. The clinical utility in genetic testing, therapy with genetically tailor-made drugs, and gene therapy is obvious. But the knowledge of these genes and their function is also important in itself as basic biological knowledge.

I have argued that scientists doing basic research should take responsibility for the applications of their findings in the sense that they should try to affect other people's use of them in an ethically acceptable direction. This holds true for geneticists studying families and ethnic groups as well. They should assume responsibility for the research, development and clinical use of genetic testing,

[13] See URL: www.stanford.edu/group/morrinst/hgdp/protocol.html

therapy with genetically designed drugs, and gene therapy. As argued earlier, two special reasons for this are that the geneticists themselves often refer to such applications in order to justify their own research, and that the new genetic knowledge often very rapidly is put to use making the applications often rather easy to anticipate.

One way to take responsibility for the application is to be a role model in taking similar ethical problems seriously in one's own research. The ethical problems in genetic family studies are to some extent similar to the problems in the clinical setting. This is particularly true of the general problem of family consent in genetic testing.

Another way of assuming responsibility for the application is to reflect on desirable goals of such application and on consequences that should be avoided, and then take part in a dialogue with other scientists, politicians, industrialists, and the general public. Through such a dialogue geneticists may come to influence the applications of their findings. The remainder of this book can be viewed as a contribution to such a reflection and dialogue regarding the applications and indirect consequences of the knowledge obtained through various kinds of gene hunting. Let us start with applications in terms of genetic testing and screening.

Responsibility for the Use of Genetic Tests and the Policy of Genetic Counseling

As a result of the HGP and the gene hunting in families and ethnic groups, more efficient and reliable methods for genetic testing are expected to be developed. As mentioned earlier, there are many different kinds of tests, for example, carrier identification, prenatal testing, preimplantation genetic diagnosis (PGD), newborn screening, testing for late-onset disorders, and DNA fingerprinting. Genetic testing includes testing for specific genes or markers as well as chromosome analysis.

Clinical geneticists in particular have a responsibility for the clinical use of genetic tests. One aspect of this responsibility is to counteract premature introduction of tests. We saw in the previous chapter that there is a risk for pressure from biotechnological companies to introduce genetic tests even when data is insufficient. For example, the data on the penetrance of the genes may be based solely on the atypical families used to identify the genes. Geneticists have a moral responsibility to counteract such pressure, although they might, of course, face a conflict of interest when they at the same time have holdings in the companies promoting the tests.

Another aspect of responsibility concerns the validity and utility even of well established genetic tests. It is important that clinical geneticists inform their patients that even such tests may have low predictive value. It is also vital that they avoid using the tests on inappropriate candidates. Not all who request a genetic test are appropriate. The probability of a positive test result or of a false negative result will be greater in some people or populations than others.

Finally, we have the problem of dialogue in using the tests. Clinical geneticists have a responsibility to ensure that genetic counseling is carried out in an ethically acceptable way. It is a part of the responsibility of the genetics community to work out a professional policy for genetic counseling. Below I shall briefly sketch a model for genetic counseling whether performed by clinical geneticists, obstetricians, or a special profession of genetic counselors. This is my main contribution in this section of the chapter. The ethical problems that will be presented are only intended as illustrations of what problems genetic counselors may face. I will give no suggestions about how to respond to them. They are best approached on a case-by-case basis. Actually, the practice of genetic counseling very clearly indicates the value of the method of imaginative casuistry.

Dialogue in Genetic Counseling: Different Models

What should characterize the professional policy on dialogue in genetic counseling: nondirectiveness or directiveness?[14] In order to answer this question certain clarifications are needed. It seems as if different terminologies are used by different participants in the discussion. Some seem to reserve the term 'nondirectiveness' for an approach according to which the counselor should only give neutral information. Others use it in a broader sense according to which nondirectiveness may also include some assistance in decision-making. I will use the term in this broader sense.

We also have to make a distinction between the issue of directiveness or nondirectiveness with regard to the content of decisions on the one hand, and psychological support in decision-making on the other. For instance, it is quite possible to combine nondirectiveness regarding the content of decisions and psychological support in decision-making. Conversely, it is quite possible to combine directiveness regarding the content with no actual psychological support in decision-making. With this in mind, it is obvious that there are several possible alternative policies. In making a classification, I find it useful to take the possible directiveness regarding the content of decision as a dividing line. Moreover, within the category of nondirective genetic counseling I make a distinction between just giving 'neutral' information and assisting in decision-making without recommending a specific content of decision. I put 'neutral' within quotation marks because those who defend this view do not necessarily believe that it is possible to provide completely neutral information. Many would argue that the goal should be to be as neutral as possible. 'Assisting', as I use the word, should be distinguished from psychological support. 'Assisting' concerns help in carrying out rational deliberation. 'Psychological support'–which, of course, is an uncontroversial demand–includes emotional and social support.

---

14 This central question is discussed in e.g. Marteau, Richards (eds.), 1996, Clarke (ed.), 1994. A clarifying overview from an international perspective can be found in Biesecker, Marteau, 1999.

With directiveness regarding the content of decision as a dividing line, the different views on genetic couseling can be summarized as follows:
1. directive genetic counseling,
2. nondirective genetic counseling:
2.1 just giving 'neutral' information,
2.2 assisting in decision-making:
2.2.1 interpreting the value commitments (weak version),
2.2.2 challenging the value commitments (strong version).[15]

In the first alternative, the goal is to help patients and prospective parents to reach a wise decision. A specific course of action is recommended in contrast to other possible alternatives. For example, prospective parents may be recommended not to undertake selective abortion on a fetus with Down's syndrome. A strength of this view is that it recognizes the need for help in decision-making. Many patients and prospective parents find it very difficult to make decisions in cases like these. They may ask about which decisions are commonly made, or want a direct suggestion by the genetic counselor. A weakness, however, is that this kind of professional policy does not take the principle of autonomy seriously enough. It is the patient or the prospective parents who should make the decision, not the genetic counselor. In addition, there are many different views among genetics providers. This means that which decision you make depends on whom you take advice from.

The main advantage of nondirective genetic counseling, on the other hand, is that it does take the principle of autonomy seriously. A problem with alternative 2.1, however, is that it does not take the need for help in decision-making seriously enough. This makes alternative 2.2 a better option. Many people want some help in these difficult decisions. The problem is how to give such help without violating the principle of respect for autonomy. Here I distinguish two possibilities that both stress the importance of assisting in decision-making. According to the weak version of alternative 2.2, the counselor should help to interpret the patient's or the prospective parents true values. It is the patient or the prospective parents who should make the decision and it is their value commitments that should be decisive. However, many people may have some problem in finding out what their true value commitments are. Here the counselor can be of great help. According to the strong version of alternative 2.2, the counselor should be a partner in dialogue and challenge the standpoints of the patients or prospective parents. He or she should give counter-arguments in order to facilitate well-considered decisions without putting pressure on the patients or prospective parents. This policy comes close to directive genetic counseling. However, the aim is not to help them reach a wise decision but to help them to reach a decision wisely by considering many different perspectives and arguments (cf. Shiloh, 1996). Actually, such consideration is precisely what moral imagination according to imaginative casuistry is all about.

---

[15] In this classification I have been inspired by Emanuel, Emanuel, 1995, where the authors describe four major models of physician-patient interaction. Cf. Tong, 1997, pp. 235-238.

In my opinion, alternative 2.2.2 is preferable. I call this the 'nondirective assistance model of genetic counseling'. This view on genetic counseling is consistent with the view on the long term goals of reprogenetic medicine that I will propose in Chapter 6. However, it should be noted that the nondirectiveness holds only within the boundaries of disease. As I shall argue, nondisease-related genetic intervention should be discouraged in medicine. According to the model, seeking genetic information not only about serious diseases but also about nonserious afflictions should be accepted in principle, while seeking genetic information about nondisease characteristics should be discouraged. By adding 'in principle' I keep the possibility open that the resources within a public health care system might not be sufficient enough to let clients seek genetic information about nonserious afflictions. Genetic testing for serious diseases must be prioritized. But, of course, drawing the line between serious and nonserious afflictions can be very difficult. In private health care, however, genetic information about nonserious afflictions should be given if the clients are prepared to pay. This they can hardly be denied.

In sum, the model proposed here implies that genetic counseling should (1) be nondirective regarding content of decision, and (2) aim at assisting the patients/clients in reaching a decision wisely by challenging their views and presenting various alternative standpoints and arguments.

Ethical Problems of Dialogue

Many kinds of ethical problems arise in genetic counseling. One may view these from different perspectives, for instance from the perspectives of the patients or prospective parents who come for genetic counseling, or from the perspectives of the genetic counselors. Below I shall focus on the ethical problems as they appear to the genetic counselors whether clinical geneticists, obstetricians, or genetic counselors as a specific profession. How the ethical problems are approached depend on what kind of model of genetic counseling one starts from. Now and then I shall briefly describe the implications of the 'nondirective assistance model'. In general, however, no substantial solutions to the problems will be given. The aim is primarily to illustrate what kinds of problems are raised. As stressed, the problems are best approached on a case-by-case basis and it is beyond the scope of this study to carry out such case discussions. Note that most of the problems are directly related to dialogue.

To Inform about Genetic Risks

A key task in genetic counseling is to inform about genetic risks. This is not easy. By their method of selecting and presenting information, counselors always influence the decision of their clients, whether they want it or not. Different ways of presenting genetic risks may result in different decisions. Completely neutral information is not possible (Lippman, Wilfond, 1992). This does not mean, however, that the goal should not be to provide as neutral information as possible.

Moreover, there are many ways in which patients/clients may misunderstand the information given. One source of misunderstanding is the use of deterministic language and metaphors. Take, for instance, the term 'presymptomatic testing'. This term is never entirely appropriate, although it may be easier to accept in some cases (for example, Huntington's disease) than in others (for example, hereditary heart diseases). By using it in describing all kinds of genetic testing counselors may contribute to the patients' misunderstanding because patients are likely to overinterpret the test results. The term 'predictive testing' is also likely to be overinterpreted. Genetic counselors must be careful with the nomenclature.

Another source of misunderstanding is the ambiguity of the term 'risk'. The term is often used in different senses by genetic counselors and their clients. The counselors only refer to the probability of a negative genetic outcome. The clients, on the other hand, often refer to severity (Shiloh, 1996, p. 91). The genetic counselor has a responsibility to ensure that the concept of risk is understood appropriately.

It is also vital to explain the difference between being a carrier of a disease gene and having the corresponding genetic disease. In autosomal recessive diseases, the person carrying a single copy of the gene is usually not affected. Moreover, in dominant diseases, it is important to make a distinction between the presence of a disease gene and the phenotypic expression of it. Often, many years elapse between the genetic test and the manifestation of the genetic disease.

The relation between test result and the possibility to prevent or treat the disease must be properly explained as well. In testing for a late-onset disease such as Huntington's disease, the presence of the gene or marker for this disease makes it highly probable that the disease will be expressed. This creates problems since no prevention or treatment is available. In addition, the time of onset and severity can be predicted only with a certain probability. In newborn screening for PKU, the presence of the gene or marker for this disease also makes its expression very probable. In this case, however, prevention and treatment is available (appropriate diet). In testing for hereditary breast cancers (BRCA 1, BRCA 2) and hereditary heart diseases, we face yet another situation. If the gene or marker is present, there is a high probability that the disease will be developed. However, something can be done for prevention. Hereditary breast cancers might be prevented with high probability by mastectomy. Hereditary heart diseases might be prevented by a proper lifestyle. However, we cannot know with certainty whether these measures will be successful.

In communication with the patients/clients metaphors will probably play an important role. The use of appropriate metaphors is therefore also essential. As we have seen, the 'crystal ball' metaphor is not so good. It is too deterministic for most cases. Juengst's proposal of a 'weather forecast' metaphor is in many cases better (see p. 109).

To Seek Genetic Information or Not?

The therapeutic gap is widening. The HGP gives us more and more information about the genetic background to different diseases but our ability to give treatment and cure is not increasing at the same pace. In a clinical context, this raises serious problems. When there is no treatment, clients may find it extremely difficult to decide whether they want a certain genetic test to be carried out.

Why would anyone request genetic testing? One reason can be that one shows certain symptoms of a genetic disease and want confirmation or disconfirmation by genetic diagnosis. Another could be a family history of disease. Other family members in previous or present generations have developed the disease. By undertaking a genetic test one wants to know whether one has the disease gene or not. If one turns out not to have it, this would be a great relief. If one turns out to have it, this information could be of help in undertaking certain preventive measures or, if no such possibilities are open, to plan for the remaining years of one's life. Important examples of diseases in which family history could be an incentive to request genetic testing are late-onset disorders such as Huntington's disease and hereditary breast cancers.

However, the frequency of requesting genetic tests does in fact vary depending on the disease. Studies show that persons with a family history of Huntington's disease have a general tendency not to request tests, while persons with a family history of hereditary breast cancers are requesting tests to a rather great extent (Marteau, Richards, 1996, pp. 3-5, 27). It is not difficult to understand this difference. In the case of Huntington's disease, there is nothing one can do to prevent the disease. To live with the knowledge that within not too distant a future one will become terminally ill is to many people a terrible option. They find it better not to know and live with at least some hope until the disease hits them. In the case of hereditary breast cancers, on the other hand, there is a possibility of preventing the disease. By undertaking mastectomy the risk of developing the disease may be to a large extent reduced. But, of course, not all women are prepared to make such a radical decision. Women at risk will develop breast cancer with only a certain probability meaning that they may not develop it at all. In this case they would have removed their breasts in vain.

A family history of disease can be an incentive to undertake genetic testing also for people who are planning to have children. This holds, of course, for diseases such as Huntington's and hereditary breast cancers but also for recessive disorders like cystic fibrosis. If both partners have a family history of cystic fibrosis, they may request genetic testing to find out whether they are carriers. If both are carriers, they may decide not to have children.

Moreover, prospective parents may under certain circumstances request prenatal testing. For example, women over 35 are at an increased risk of having children with Down's syndrome. If the fetus turns out not to have Down's, this would be a relief to many prospective parents. If it turns out to have Down's, this information could be a reason for some prospective parents to terminate the pregnancy. Other prospective parents would choose to have the child born and would have time to prepare themselves and make plans for the future.

Some prospective parents may request preimplantation genetic diagnosis. One example is when one of them has a recessive disorder. On the basis of preimplantation genetic diagnosis, fertilized eggs that do not carry the gene can be implanted while the rest is discarded.

In sum, for many clients it is very difficult to decide whether to seek genetic information. When there is no treatment, do they truly want to know whether they themselves (carrier identification, testing for late-onset disorders) or their fetus (prenatal testing) or child (neonatal testing) have a certain clinical genotype? Moreover, when the disease in question is such that there is no way to know for sure that a disease will develop, only that there is a certain risk, should they request a genetic test? Are they better or worse off knowing? These are very difficult decisions to make and the genetic counselor has a delicate task in assisting the clients. The problem of the clients to know or not to know raises the professional and ethical problem for the genetic counselors whether to inform or not to inform.

This in its turn raises the fundamental ethical problem whether the genetic counselor should be directive. I have argued for a model of nondirective assistance. This model implies that the genetic counselor should assist in decision-making without being directive regarding the content of the decisions. How can this be carried out in practice? For instance, how should the following questions be answered: What do people usually do? What would you have done? This last question has been called 'the terrible question' (Karp, 1983). On the proposed model, it is acceptable to answer these questions, but the model stresses the importance of challenging the view of people in general, the view of the genetic counselor himself or herself, as well as the view of the client.

Moreover, what does fair treatment of clients means more precisely? A reasonable view could be that the genetic counselor should be nondiscriminating and treat everyone equally. If a genetic test is available, then it should be available to all those who request it provided that the resources are sufficient and that they are appropriate candidates. The latter is an important condition to add, because the probability of a positive test result or of a false negative result will be greater in some people or populations than others.

A third problem concerns conflicts within families regarding testing for, for example, Huntington's disease. The son may want to take the test, while the father does not want to know. If the son's request is accepted and the result is positive, then it is clear that the father will also manifest the disease. The nondirective assistance model implies that the difficulties should be pointed out but that each individual has to make up his own mind.

Finally, there is the problem of prospective parents who seek genetic information about possible nonserious afflictions of their embryo or fetus or about some of its nondisease characteristics. Should the genetics provider assist in such prenatal testing? I have already indicated that the nondirective assistance model holds within the boundaries of disease. This means that the genetics provider should assist them in seeking genetic information about diseases including nonserious afflictions but not regarding nondisease traits. Genetic information regarding nonserious afflictions should be sought only if the public health care resources are sufficient for this kind of testing or the prospective parents have sufficient personal resources to pay for the testing (see p. 252).

Genetic Information which Has Not Been Requested

Sometimes genetic information that has not been actively sought as a response to the clients' requests is still discovered by the clinical geneticist. The information can be about, for instance, a genetic disposition for another disease or misindentified paternity. How should the clinical geneticist act in such cases? This is not too uncommon an ethical problem.

Let us focus on two examples from prenatal testing. When a genetic disposition for another disease is discovered in prenatal testing this might have very troubling consequences for the clinical geneticist. Should the information be revealed to the prospective parents? When there is a treatment for the disease or a possibility of prevention there is hardly a problem. But when there is no treatment and the only alternative is to terminate the pregnancy it is not easy to know what to do, particularly when the disorder is rather severe but not severe. An example is when one is testing for Down's syndrome but finds a genetic disposition for Klinefelter's syndrome. I will return to the problem of selective abortion in such cases below. The issue that troubles us here is whether the clinical geneticist should inform the prospective parents. The nondirective assistance model implies that the prospective parents should be informed. Otherwise, it will not be they who decide but the counselor.

Another example is the discovery of misidentified paternity in prenatal testing. Let us assume that two prospective parents are both carriers of cystic fibrosis and that the specific mutations were identified in an earlier test. In prenatal testing it is now discovered that the fetus has two copies of the disease gene but that one of the mutations are different from that of either parents. This makes it extremely probable that the man is not the biological father. This raises several ethical questions: Should the clinical geneticist give complete information to the prospective parents about the results or only certain parts of it? Should the geneticist give different information to them, for instance, tell the woman that her husband is not the biological father but not tell the husband? Moreover, should the biological father be contacted and informed that there is a risk that his future children may inherit cystic fibrosis?

These are all difficult questions. My standpoint is that in cases like these the mother, and only the mother, should be informed. It must be her decision whether to inform the others. However, I will refrain from arguing for this view here. The aim of pointing out these problems is to illustrate what issues genetic counselors may face.

Informing Other Family Members?

Genetic information may concern other family members as well. In genetic linkage analysis, the cooperation of other family members is essential. In other kinds of genetic testing, information is sometimes obtained about relatives who have not asked for it. This information can be about disease but also about, for example, misidentified paternity. In all these cases the genetic counselor has a moral responsibility not only to the client but also to the other family members.

In genetic linkage analysis, the geneticist has to find a way to recruit other family members. Earlier in this chapter I discussed this issue in the research context. However, in the clinical context the task may actually be somewhat simpler. The other family members may have something to gain. However, if some family members refuse to participate in the analysis, should they, in order to help the client, be put under pressure to do so?

Obtaining genetic information about other family members who have not requested it also raises difficult problems. Should they be informed? In matters of disease, it depends on the answer of the client. If the client says 'no', this should, as a general rule, be respected. If the client says 'yes', it depends on the kind of disease and other circumstances.

Information about misidentified paternity should be handled with care. As indicated above, in prenatal testing it must be up to the mother to decide whether to reveal it. In other kinds of genetic testing, it should not be mentioned at all.

Selective Abortion and Preimplantation Selection

Genetic information about embryos and fetuses can be used as a basis for selection procedures.[16] Preimplantation genetic diagnosis (PGD), i.e. genetic diagnosis on

---

[16] See, for instance, Munthe, 1996, and Munthe, 1999.

fertilized eggs in the process of *in vitro* fertilization, may lead to embryo selection. Prenatal testing may be followed by selective abortion. These are very controversial issues. Some prospective parents would neither accept preimplantation selection nor selective abortion. Some would accept the one but not the other. Others again would accept both. Among those who accept at least one kind of selection there are also different views on which kinds of diseases or nondisease traits it would be acceptable to select against. Some people would say that selection is acceptable only regarding very severe and severe diseases, some also regarding less serious diseases, minor afflictions, disease susceptibilities, or late-onset disorders. Others would accept selection against certain nondisease traits, for instance sex, as well.

What should be the policy of geneticists regarding preimplantation selection and selective abortion? I shall discuss this controversial question from a broader perspective in Chapter 6.

To Inform about Psychosocial Risks

Before a genetic test is undertaken the patient should be informed about the psychosocial risks. The genetic information obtained by the test might be of interest to employers and insurance companies. There is a risk that patients or clients will be discriminated against (see pp. 160-165).

Genetic testing might also create problems in the family. The genetic information obtained often has implications for other family members also. Some might want the information, while others do not. In particular, of course, information about misidentified paternity may have very troubling consequences. The genetic counselor should inform about these risks before a genetic test is undertaken.

Conclusion

Moral responsibility for applications of human genetics research implies responsibility for the development and use of genetic testing. A particularly important aspect of genetic testing is genetic counseling. It is suggested that this kind of dialogue should be nondirective with regard to the content of decision, i.e. the clients should decide autonomously how to proceed. This does not exclude, however, some assistance in rational deliberation. Moreover, the provision of genetic information in genetic counseling is a very delicate task that requires a casuistic approach.

5. GENETIC SCREENING AND PUBLIC HEALTH WORK

Commonly, the ethical implications of genetics have been discussed from a clinical perspective. However, they also have to be discussed from the perspective of public health work. This aspect has often been neglected in the ELSI research.[17] What significance does genetics have for public health work? What role can and should genetics play in public health work? Genetics provides new tools. How should these be used? Which applications are ethically acceptable? It is vital that responsible geneticists use their moral imagination to find answers to these questions. It is essential that they do this in full awareness of historical cases, most importantly the eugenics movements in the first decades of the 20th century.

Traditional Public Health Work

Public health work includes all measures aiming at health promotion and disease prevention. The traditional public health work uses direct as well as indirect methods. The former includes direct treatment of disease. The latter can be a matter of giving health information in order to influence people's lifestyle or changing the physical or social environment in order to create possibilities for a healthy lifestyle. The preventive and health promoting measures may be applied at different levels: at the population level, at group level, or at the individual level.

With regard to prevention, one usually distinguishes three different kinds. Primary prevention in terms of general and specific health promoting measures can be undertaken even when no symptoms are shown. Secondary prevention can be carried out when limited symptoms are shown, and consists of early diagnosis and treatment. Tertiary prevention is accomplished when more comprehensive symptoms are shown. Here the aim is to prevent complications that are disabling.

The Use of Genetics in Public Health Work

It is obvious that the new genetic knowledge can play an important role in the health promoting and preventive work. Genetic screening and testing implies new possibilities for predictive medicine, and is, therefore, of importance not least for the primary prevention at different levels.

Of particular importance for public health work is genetic screening at population level. Genetic screening can be defined as systematic genetic testing of a population or group in order to early discover or exclude genetic disease, genetic susceptibility, or genetic carriership (Dierickx, 1999; Chadwick, 1999). There are several different types of genetic screening: prenatal screening, neonatal screening, screening for late-onset disorders, and carrier screening. Regarding more and more diseases, and also diseases that are widely spread in the population, the genetic risk can be detected long before any symptoms are shown.[18] After genetic screening, we

---

[17] However, an example of a discussion of these issues can be found in Khoury et al. (eds.), 1998.

[18] Note that I am not talking about carriership here.

may in some cases provide tailor-made prevention in terms of health promoting measures or treatment for individual persons' specific genotypes.

Often, however, there is nothing to do. This means that predictive medicine not only creates new possibilities for prevention. It also creates new problems. Some problems are related to dialogue. One problem of dialogue concerns health information and disease information. Shall we actively seek and give genetic information that the recipients perhaps do not want? We have already met this problem in the clinical context. It is exacerbated in the public health context.

Another problem of dialogue concerns how general the information about health promoting and preventive measures should be. People's tendency to develop certain diseases varies depending on genetic factors. Some may lead a very long life in spite of an unhealthy lifestyle. Others may die at a very low age because of, e.g. heart disease, in spite of a very healthy lifestyle. With this in mind, to what extent can it be justified to propose general health recommendations regarding lifestyle, food habits etc.? Can this even be misleading?

A very important question is how actively genetic medicine should look up individuals at risk. The question is important, partly because of the tension between public interest and individual interest, partly because of the therapeutic gap. The active search for people is particularly loaded when clinical phenotypes are linked to specific ethnic groups.

In the discussion of the relevance of genetics for public health work, it is important to make a distinction between phenotypic and genotypic prevention.[19] I suggest the following two stipulations:

> 'Phenotypic prevention' = def. an effort to prevent the expression of genetic disease in an individual.
>
> 'Genotypic prevention' = def. an effort to prevent the birth of an individual with a particular clinical genotype.

Phenotypic prevention is an undisputed goal in public health work. Genotypic prevention, however, is very controversial. Eric Juengst stresses that these two concepts have different histories, are based on different assumptions and have different degrees of moral authority. In his analysis of the discussion on germline gene therapy, he has found that the participators often conflate these two meanings of the term 'prevention'. They start with talking about phenotypic prevention but slide into talking about genotypic prevention (Juengst, 1995b). Germline gene therapy, however, is a method of genotypic prevention that is not available at present and will probably not be so for a long time. Selective abortion and, to some extent, preimplantation selection are the only methods that can be used at present.

---

[19] The distinction has been suggested in Juengst, 1995b.

Possible methods for phenotypic prevention, on the other hand, are, for instance, adjusted health information in order to change lifestyle, modification of the physical and social environment in order to create possibilities for a healthy lifestyle, ordinary drug therapy, therapy with genetically tailor-made drugs, somatic gene therapy, and, possibly in the future, germline gene therapy.

The distinction between phenotypic and genotypic prevention is vital, not least from a historical perspective. Genotypic prevention constituted a central part of the eugenics in the first decades of the 20th century in, for example, Nazi Germany, the US (Kevles, 1995 (1985)), and Scandinavia (Broberg, Roll-Hansen, 1996). These cases must be taken into consideration of responsible geneticists.

Are there any modern cases of genotypic prevention? A recent case is the beta-thalassaemia screening program in Cyprus in 1972-1983. During a first phase, marriages between couples in which both persons are carriers were discouraged. During a second phase, antenatal diagnosis was introduced, making it possible for carrier couples to have children free of beta-thalassaemia. Moreover, since 1983 the Church has required a premarital certificate indicating that carrier testing was done. As a result of this process of education and sensitization together with social pressure, practically no new thalassaemic children have been born in Cyprus since the beginning of the 1980s.[20]

We fiind another case in the proposal put forward by the European Commission in 1988 of European genome project entitled 'Predictive Medicine: Human Genome Analysis'. I have already referred to this proposal in the previous chapter; let me repeat its key aspects. The proposal was called a 'health measure'. Its aim was summarized as follows: "Predictive Medicine seeks to protect individuals from the kinds of illnesses to which they are genetically most vulnerable and, where appropriate, to prevent the transmission of the genetic susceptibilities to the next generation."[21] The program was criticized for being an enterprise in preventive medicine and including eugenic ideas. The protection of people from transmitting genetic diseases or conditions would almost always involve eugenic decisions about what is normal and abnormal, desirable and undesirable. As a response, the European Commission submitted a new proposal that did not mention predictive medicine, simply called 'Human Genome Analysis', and this proposal was formally accepted by the European Community Council of Ministers in 1990.[22]

This presentation of genetics in public health work should be viewed as introductory. I will discuss genotypic prevention and eugenics extensively in Chapter 6. But let me make it clear that I am critical to genotypic prevention as a public health measure. Genetics should be used in public health work only for phenotypic prevention. I will also discuss the tension between a public health

---

[20] For a discussion, see Hoedemaekers, ten Have, 1998.

[21] Commission of the European Communities, 1988. Quoted in Kevles, 1994, p. 20.

[22] See Commission of the European Communities, 1989; European Community, 1989; Official Journal of the European Communities, 1990; Kevles, 1994.

framework and a client-centered framework. This tension becomes particularly obvious in relation to genetics. In cases of conflict, should the goal be reduction of the incidence of genetic disease in the population or increased reproductive autonomy of prospective parents?

6. GENETIC INFORMATION IN EMPLOYMENT AND INSURANCE

I have stressed that genetic researchers as well as genetic counselors in their dialogue with patients should inform about psychosocial risks related to seeking genetic information about diseases. The information can be very interesting for insurance companies and perhaps also to employers, and they may use it in a way contrary to the interests of the research subject or client. In line with the general view of responsibility in science presented earlier, geneticists have a moral responsibility to try to affect the use of genetic information in employment and insurance in an ethically acceptable way. Below, some arguments are presented which responsible geneticists should take into consideration and explore by means of moral imagination.

Genetics at the Workplace

It is often stated that knowledge of a person's genetic make-up may be of great value to the employer. The company may, for instance, want to refuse to employ people at risk for certain diseases with genetic components. This might, of course, lead to discrimination of such people (Murray, 1992, pp. 62-66; Asch, 1996). However, there are actually very few examples of employers who have used testing. In addition, there are very few tests in which the results are likely to predict with sufficient accuracy the employee's ability to perform the job and perform it without danger to himself or herself or to others.

Moreover, a more attractive way of handling a situation with potential employees at risk would be to change the environment so that they can nevertheless work there. This can probably be done in many cases. Sometimes, however, it may not be possible, and then it might be justified not to give them that particular job. A problem is, of course, what it means to be at risk and what risks are acceptable. What we are talking about here are persons who have not developed a certain disease but have a genetic susceptibility of doing so in a particular environment.

However, there is an important debate on the issue of whether employers should at all be allowed to require or request genetic information in the first place. It is vital that the genetics community takes part in this debate. Some argue that legal prohibition of employers to require or request genetic information is the only way to avoid discrimination. Others doubt that this would work in practice. Employers might simply give the jobs to applicants who on their own initiative provide the genetic information.

It seems necessary, however, with some legal prohibition of employers to require or request genetic information. There may be some kinds of jobs regarding which certain employers may find reason to request or even require genetic information, and there is a risk for discrimination. The prohibition should not, however, be completely general. Certain exceptions could perhaps be justified. Moreover, in most cases a legal prohibition does not seem necessary. Most applicants would be anxious to take care of themselves and would not take a particular job if there was a high risk that they would develop a disease. They would be prepared to disclose relevant genetic information about themselves to the employers and even undertake a genetic test. But, of course, there is still a possibility that some people who are unemployed may be prepared to take risks.

Alongside the risk that employers will discriminate against people because of directly required or requested genetic information, there is a more urgent problem. They may not want to employ certain people because it is not possible to insure them. This problem is already with us independently of genetics. People may not be allowed to get any insurance due to nongenetic information. As a consequence of genetics, it may occur that they do not get any insurance due to genetic information. This brings us to the more general issue of the implications of genetics for insurance.

Genetics in Health Insurance and Life Insurance

Insurance companies may also want genetic information about their applicants. This might create serious problems (Murray, 1992, pp. 66-71; Stone, 1996). If the applicant alone has this information, he or she has some advantage compared to the insurance company. With regard to life insurance, the applicant's family may obtain much more insurance money than it normally would. With regard to health insurance, the applicant may get more extensive health coverage than he or she (or the employer) has paid for. On the other hand, if the insurance company has the information, the applicant may have to pay very high premiums or perhaps not be allowed to get any life insurance whatsoever, or be allowed to buy only limited health coverage or denied health coverage all together (Capron, 1990, pp. 689-696).

A key notion in insurance is underwriting. It is the process by which insurance companies determine whether an application for insurance represents an acceptable risk. This determination is based on the kind of contract that the applicant desires and on certain characteristics of the applicant. The potential health impact of these characteristics is compared with other applicants, and then the risk is classified as preferred, standard, substandard, or declined. Traditionally, the characteristics relevant to underwriting have included factors such as age, gender, health history, physical condition, occupation, the use of alcohol and tobacco, family history of disease, and serum cholesterol. On the basis of the underwriting, the premiums are set according to the principle of equity, i.e. the premiums vary by risk. This principle should be distinguished from that of equality, according to which everyone would pay the same price (Pokorski, 1994, pp. 92-93).

The problem raised by modern genetics is whether insurance companies in the process of underwriting should be allowed to require or request genetic test results. In discussing this problem, it is important to distinguish between different kinds of insurance such as health insurance and life insurance. It is also important to be aware of the context. Are we talking about countries with a well developed public health care system, such as Canada and the European countries, or countries in which private health care dominates, such as the US?

The use of genetic information in insurance is very controversial. Some countries have legally prohibited insurance companies to require or request genetic test results. Some states in the US also have such laws.[23]

I would like to put forward the following proposals. They are only intended as a contribution to an interdisciplinary dialogue. Let us start with health insurance. The best way to solve the problem of the use of genetic information in insurance seems to be a mixed system with a public health insurance that may be supplemented with private health insurance for those who want it and can afford it. The public health insurance provides access for all to basic health services and some compensation for lost income during the disease period. The private health insurance may include access to special medical treatment and care or provide extra compensation for lost income during the disease period in addition to the compensation given by the public insurance. This kind of a mixed system was also the solution suggested in 1993 by the US Task Force on Genetic Information and Health Insurance with Thomas Murray as chair (ELSI, 1993, pp. 9-10). Moreover, it is the system already established–in different versions–in Canada and the European countries. It is obvious that with a well-functioning public health insurance system the interest of individuals for additional private health insurance would be reduced. Some people, however, would want such additional insurance.

The reasons for this mixed system are strong. Public health care spreads the risk to all tax-paying citizens and gives each citizen a good basic health insurance. This is in line with the principle of justice. On the other hand, those who are prepared to pay for a supplementary health insurance can hardly be denied to do this. This is required by the principle of respect for individual autonomy.

It is important, though, that insurers not be allowed to require that genetic testing be carried out. They should only be permitted to request information of results of genetic tests that have already been carried out. A request would be in line with the tradition of underwriting. To require that genetic testing is carried out would be to go beyond established practice. Moreover, insurance companies should be allowed to request such genetic information only for health insurance above a certain limit. This would imply somewhat higher premiums for all policy holders but this would

[23] For an overview of the American laws, see Frankel (ed.), 1997, pp. 103-107.

be an act of solidarity, which most people would likely find quite acceptable. To decide what the limit should be is, of course, a delicate matter that requires a balancing of many different factors.

In general, however, the supplementary private health insurance system requires that the insurance companies have the same knowledge, including genetic knowledge, as the applicants. The argument is that not having the same knowledge would be unfair to the insurers. It would undermine the basis for all insurance work, the very philosophy of underwriting. It may also make it difficult for the insurance company to stay on the market. The company may in the end suffer serious economic losses, although it for some time may be able to compensate with higher premiums for all policy holders. If we accept a private insurance market, it seems necessary that the insurance companies get the same knowledge including genetic knowledge as the applicants, at least for health insurance above a certain limit.

Much more important, however, is that not having the same knowledge as the applicants would be unfair to other applicants. If applicants would get more health coverage than they should have got, if the genetic risks were known to the insurance companies, then the price of coverage would rise for all policy holders (ELSI, 1993, p. 8). This would be acceptable for health insurance under a certain limit but not above that limit.

Moreover, as Robert Pokorski stresses, there may actually be fewer rejections if the insurance companies obtain genetic information, as this may help them to assess the risks more precisely (Pokorski, 1994, p. 104).

Finally, it is probable that it will be increasingly difficult to keep genetic information separate from other medical information that the insurance companies already have access to. It might be possible to ask physicians to keep these different kinds of information separate but much of the information about risk factors, diseases, diagnostic tests and treatments will in fact reveal genetic information (ELSI, 1993, p. 8). In other words, there are practical difficulties with a policy allowing insurance companies to use nongenetic information but prohibiting them from using genetic information.

An objection to the idea that insurance companies should be allowed to request genetic information is that applicants with certain clinical genotypes would be denied health coverage completely or only be allowed to buy a limited coverage, and this is not fair. They would be discriminated against. However, with a mixed system where basic health services are provided to all, the bad consequences would be radically reduced.

Another objection is that if insurance companies get hold on genetic information, this would threaten confidentiality. The confidentiality of genetic information is even more important than other medical information, because genetic information about an individual is also information about that person's family. However, history shows that the insurance companies have protected the confidentiality of medical information very well (Pokorski, 1994, p. 105). There seems to be no reason to believe that they will not protect the confidentiality of the genetic information.

A final remark: a mixed system in which the insurers do not request genetic information under a certain limit does not seem to require legal regulation. In Sweden, for example, an agreement between the government and the insurance companies' national organization has been quite sufficient.

If we turn to life insurance, a similar mixed system is to be preferred. The state provides a kind of basic life insurance in terms of financial support to the children of the deceased as well as to widows and widowers. Due to limited resources, however, this support would probably be more limited than the public health insurance. This support can be supplemented by private life insurance. But since the public life insurance would be more limited than the public health insurance, supplementary private life insurance would probably be more common among the general public than private health insurance. As was argued with regard to health insurance, insurers should not be allowed to require that genetic testing be carried out, only to request information of results of genetic tests that have already been carried out. Moreover, they should be allowed to request such information for life insurance above a certain limit. In addition, as was the case with health insurance above, such a system does not seem to require legal regulation. An agreement between the government and the insurers would be sufficient.

This mixed system can be justified in the same way as the mixed system with regard to health insurance, i.e. by reference to the principle of justice and the principle of autonomy. It spreads the risk to all tax-paying citizens and gives all a least some basic life insurance. Moreover, it respects the autonomy of those who want to buy additional insurance.

However, supplementary private life insurance requires that the insurance companies have the same information, including genetic information, as the applicants, at least for insurance above a certain limit. Below a certain limit people could be allowed to purchase life insurance without disclosing genetic information. This seems reasonably fair to insurers as well as to other applicants. The economic losses for the insurance companies would be negligible, and although all applicants would have to pay higher premiums in order to compensate, this would also be acceptable, since the mixed system has the advantage that most applicants would not have to disclose their genetic profile. Family members of all deceased persons obtain from the state a basic life insurance in terms of at least some financial support. The additional private life insurance could be rather limited for most people.

In conclusion, genetics may give rise to difficult problems in insurance. The best way to handle these problems seems to be a mixed system where the state gives all citizens basic health services and provides at least some financial support to the families of deceased people. Supplementary private health and life insurance should be open to those who can afford it. With regard to health insurance, the insurance companies should be allowed to request genetic test results only above a certain limit. This holds true also with regard to life insurance. It is essential that the genetics community takes these problems seriously. In order to try to affect the development, they should work out a common professional policy on these issues and participate in a dialogue with the government and the insurance companies.

7. GENETICALLY TAILOR-MADE DRUGS

Another important application of the genetic knowledge obtained through the HGP and the gene hunting in families and ethnic groups is genetically tailor-made drugs. A new branch of genetics research is emerging called 'pharmacogenomics'. It should be distinguished from pharmacogenetics, which is "the study of the linkage between an individual's genotype and that individual's ability to metabolize a foreign compound." Pharmacogenomics involves "the application of genomic technologies such as gene sequencing, statistical genetics, and gene expression analysis to drugs in clinical development and trials" (Emilien et al., 2000, p. 393). This kind of research aims at understanding at the genomic level why individuals often respond differently to drug therapy.

Preclinical and Clinical Research

Pharmacogenomics may be applied in preclinical as well as clinical research. In preclinical research, one option is to start screening for compounds with the least variation between individuals. When a particular gene is selected as a target, the compound that works best compared to its subtypes may be chosen for further research. Pharmacogenomics may also be used to select out adverse effects before drugs enter the clinic. For example, the gene-expression pattern for the liver of an animal administered a drug can indicate whether gene pathways related to toxicity have been turned on (Emilien et al., 2000, p. 394).

Pharmacogenomics may be applied in the clinical setting leading to smaller and faster clinical trials with regard to efficacy and safety. In clinical trials and clinical practice, the key idea is to identify and select patients with genotypes that are expected to be the most suited to treatment with a particular drug. The drug can be an existing drug or a drug under development. The various responses of different patients to a drug depend at a fundamental level on their genetic constitution. In relation to a specific drug, there may be roughly three different categories of patients:

(1) responders,

(2) nonresponders,

(3) patients for whom it is 'not safe to prescribe' the drug.

Only individuals categorized as responders will take part in randomized clinical trials of the drug. Nonresponders will be used to find other known compounds to which they respond. In this way the drug can be re-engineered to suit them. Patients of category (3) will be used in the search for new substances. A new drug may be developed which will not have toxic effects on these patients. It should be noted, however, that the three categories seldom are clear-cut.

Moral Imagination: Arguments For and Against

What should responsible geneticists consider with regard to pharmacogenomics? As with other kinds of genetics research, they must use their moral imagination and envision a variety of arguments and problems.

Different kinds of arguments have been put forward in favor of pharmacogenomics. The main *pro* argument is, of course, that better and new drugs can be developed. Increased efficacy and safety of the drugs can be achieved. This improves health care by increasing the quality of life of the patients and minimizing their suffering. It also reduces health care costs.

Other arguments concern certain advantages in the research process. The number of patients to be selected for clinical trials is reduced, and the time for clinical trial is reduced. Nonresponders and 'not safe to prescribe' patients can be excluded from clinical trial. This means that the total amount of human suffering is reduced. The 'not safe to prescribe' patients are not used in the study at all, and can, consequently, not be intoxicated. Responders in the control group who do not get the drug will also benefit since the drug will be available to them more rapidly. Another advantage is a reduced number of animal trials. This means that the total amount of animal suffering is reduced.

The option of individualized treatment seems very attractive. Pharmacogenomics is, however, a still emerging field and not so much research has as yet been carried out. We will have to wait and see whether the expectations hold what they promise. Moreover, the advantages with regard to the research process also seem very positive. I will discuss this more extensively below.

A problem with this kind of drug research regards genetic privacy. It is obvious that the genes of individual patients are exposed in the research process itself. However, this problem is similar to problems raised by genetic testing and I shall not discuss it here.

Other problems are raised by the commercial aspects. I have dealt with the commercialization of genetics in the previous chapter. This commercialization becomes very obvious in pharmacogenomics, perhaps more so than in any other field of genetics. But I shall not discuss the commercialization any further, either.

Another problem is that the genetically tailor-made drugs will probably be very expensive, at least initially. The reason for this is that genetically tailor-made drugs will be designed for smaller groups of patients, instead of larger groups. This may create financial problems in public health systems. It may be that the health care costs are reduced in the long run with more efficient drugs but in a public health care system there may occur financial problems in the short run.

I will not go deeper into these problems. They cannot be ignored, but seem possible to handle. On balance, the advantages of pharmacogenomics seem to outweigh the disadvantages.

Instead I shall focus more in detail on ethical problems raised by randomized clinical trials in pharmacogenomics. Randomized clinical trials have been subject to much ethical discussion. But, as stressed in the argument above, such trials will get a somewhat different character in the case of genetically tailor-made drugs as compared to conventional drug design. This difference may be ethically relevant and should be taken into consideration by the responsible geneticists.

Let me first make a brief analysis of the ethical aspects of randomized clinical trials in general.

Randomized Clinical Trials (RCT)

To show that an experimental drug is the cause of recovery, one must compare the treatment using that drug with treatments using conventional drugs or a placebo. In order to achieve this, the research subjects are divided in two groups, an experimental group and a control group. The experimental group gets the experimental drug and the control group a conventional drug or a placebo.[24] The hope is that the experimental group gets a higher rate of cure than the control group and that this result is statistically significant. The result is statistically significant when one can rule out the possibility that the difference is the outcome of chance. However, there is an additional problem. One must also rule out that the difference is caused by some other factor such as the seriousness of the illness. To solve this problem randomization is used. The research subjects are assigned to the experimental group and the control group by a random selection process. Because the influence of other factors can be assumed to be equally distributed within each group, the difference between the two groups will depend on the experimental drug.

However, the fact that the research subjects are patients creates a serious ethical problem. It has been labeled 'the RCT dilemma'. The one horn of the dilemma is the health of the patients, the other is scientific significance. The dilemma emerges when the evidence is approaching, but has not yet reached, the level of significance. At some point during the trial, however, we draw the same conclusion concerning the efficiency of the experimental drug as we will do at the end of the trial (for instance, 95%) but at a lower level of confidence (for instance, 70%). This level of confidence may not be high enough for publication. But it would perhaps be high

---

[24] Note that according to the *Helsinki Declaration* (the revised version of 2000) patients in the control group should be provided "the best current prophylactic, diagnostic, and therapeutic methods." But the Declaration also stresses that this does not exclude the use of placebo when no such proven method exists. See URL: www.wma.net.

enough for treating patients, in particular when the disease is very serious and there are no other options (Gifford, 1994).

The dilemma can also be formulated in terms of responsibility. As a scientist, the researcher has a responsibility to demonstrate scientifically significant results. As a physician, he or she has a responsibility to promote the health of the patients. Let us call the former responsibility 'the scientific responsibility' and the latter 'the therapeutic responsibility'.

There has been a great deal of discussion about the RCT dilemma. Several different solutions have been put forward, focusing on different horns of the dilemma (Gifford, 1994).

Some argue that the dilemma can be avoided. One suggestion is that the therapeutic responsibility is the only responsibility of the researchers. The important thing is to help people and to the extent that RCTs do not, they are not acceptable. Another suggestion is that the scientific responsibility is the only responsibility. The task is to obtain scientifically reliable results. The researcher has no therapeutic responsibility. A third possibility could be to use other methods for clinical trials.

According to another kind of solution, the RCT dilemma cannot be avoided. A violation of the therapeutic responsibility is inescapable but, it is argued, this violation can be justified. A utilitarian argument is that denying the benefit of optimal treatment to the research subjects will generate greater benefits for future patients. A contractarian argument is that since each of us benefits from social ventures such as drug development, each of us has an obligation to accept some costs or risks in particular cases. Other arguments stress informed consent and altruism. It should be left to the patients to decide whether to participate in the RCT after receiving information about randomization, risks, and placebo or nonoptimal treatment in the control group. If they autonomously decide to participate, this is quite acceptable. It is particularly acceptable when patients choose to participate due to an altruistic willingness to help others. There is also an argument that if the omission of giving optimal treatment cannot be avoided, the harm should at least be minimized.

The dilemma is real and cannot be avoided in clinical research. The therapeutic responsibility as well as the scientific responsibility should be taken seriously. In many cases there are no real alternatives to RCTs. The therapeutic dilemma is something we have to live with in clinical research. Whether the therapeutic responsibility outweighs the scientific responsibility depends on the context. The dilemma is contextualized. RCTs are more or less acceptable. Sometimes, the costs are too high. This is particularly obvious in cases of terminal illness, as has recently been argued by Udo Schüklenk (Schüklenk, 1998). But sometimes the harms are acceptable; the value of achieving scientific reliability outweighs the harm. This value is due to the public good, i.e. the great benefits to future patients. In these cases the important thing is to minimize the harm while still demonstrating the scientific reliability. However, in these RCTs informed consent is indispensable.

This is not recognized in the utilitarian and contractarian arguments, which both place too heavy a burden on patients in the control groups. If they do not want to participate they should not be required to do it. On the other hand, if they are voluntarily prepared to participate for altruistic reasons after open dialogue on risks, etc., they should be allowed to do so. This altruism should be gratefully accepted by the researchers as an important resource in research.[25]

Randomized Clinical Trials with Genotypically Selected Patients

In a review article, it is anticipated that the "pharmacogenomic developments should lead to smaller, more rapid and cost-effective trials, and ultimately to more individually focused and effective therapeutics" (Emilien et al., 2000, p. 395). What are the implications of this for the ethical RCT dilemma?

On the one hand, it seems as if that the dilemma is exacerbated in the sense that it emerges at an earlier point of time. Only individuals roughly belonging to the category of responders to the drug participate in the clinical trials. This means that we already from the start of the trials have good reason to believe that the drug has some positive effects on these patients. The only thing is that we do not know exactly how positive. This holds for efficacy as well as safety.

On the other hand, the dilemma is more easily solved in this kind of research than in traditional clinical trials of drugs. The time for clinical trial is reduced. The number of patients participating is also reduced. Moreover, the expected benefits are greater, because we have good reason to believe, already in advance, that the patients will respond. Finally, it will be possible to give the drug more rapidly to the patients in the control group as well. All this together makes it easier to maintain that the scientific goal of getting reliable data outweighs the therapeutic obligations to the participating patients.

Conclusion

Pharmacogenomics is a rapidly emerging field and it is essential that responsible geneticists take into consideration the ethical aspects sooner rather than later. A dialogue between geneticists, industrialists, politicians, and the general public is essential. Critics object to the commercialization of genetics, which is particularly obvious in this kind of research. On the other hand, pharmacogenomics opens the possibility of genetically tailor-made drugs, which would be of great medical value. It is particularly important to recognize that from the point of view of research ethics, this kind of research gives randomized clinical trials a new character.

---

[25] For a discussion of altruism in health care and medical research, see Nordgren, Westrin (eds.), 1998.

CHAPTER FIVE

GENETIC MODIFICATION OF HUMANS AND LABORATORY ANIMALS

1. DIALOGUE AND GENE THERAPY

The Human Genome Project produces knowledge of relevance to human gene therapy research. This means that gene therapy can be viewed to some extent as an application of results from this project even if the research on gene therapy started before any results were published.

Gene therapy research raises many 'ordinary' issues in research ethics, for instance, choice of disease and dialogue with patients. Ultimately, however, it is the potential in the long term of gene therapy that has to be considered. Responsible geneticists have to take part in a dialogue with the general public and politicians on a number of important issues: Which applications of gene technology to humans are morally defensible? To what extent is it morally acceptable to modify the human genome? Are there limits that should never be transcended? At present, somatic cell gene therapy is the only kind of gene therapy that is actually tested. But what about germline gene therapy? Should scientists be permitted to carry out germline gene therapy research? This is perhaps the most controversial ethical issue in human genetics. Germline gene therapy is contrary to, for example, the Convention of Human Rights and Biomedicine issued by the Council of Europe (Council of Europe, 1996). Moreover, in a report issued by the American Association for the Advancement of Science (AAAS, publisher of *Science*) a total moratorium is called for.[1] On the other hand, a few important geneticists have recently suggested that germline gene therapy research should be started.[2]

With this in mind, more space will be devoted to germline gene therapy than to any other specific issue discussed in this book. This might give the book a slightly unbalanced character. It reflects, however, the extra-ordinary controversiality and importance of the issue. Another controversial issue is genetic enhancement, whether somatic or germline. Ethical problems raised by this issue will be discussed as well.

---

[1] See URL: www.aaas.org/spp/dspp/sfrl/germline/main.htm

[2] See URL: www.ess.ucla.edu:80/huge/report.html.

However, genetic modification of animals has been carried out long before any genetic modification of humans. In addition, transgenic animals are used in human genetics research, for instance, as disease models. This makes it necessary to discuss ethical problems raised by the genetic modification of animals as well, despite the fact that this book is about human genetics. The final section of the chapter will be devoted to such a discussion.

Different Kinds of Genetic Modification

Gene therapy can be defined as a set of approaches to the treatment of human disease based on transfer of genetic material (DNA) into an individual. Transfer may be achieved in vivo or in vitro. In the former case, gene delivery is made directly to blood or tissue, in the latter case indirectly through the introduction of cells modified in the laboratory to harbor foreign DNA. One important method for the transfer is to use viral vectors, but there are also other methods (Smith, 1999; Brenner, 1999).

In 1980, the first unauthorized experiment of somatic gene therapy was performed (Cline). Ten years later the first officially authorized experiment was carried out (Rosenberg et al.) (cf. Walters, Palmer, 1997, pp. 17-36). At present, several thousand patients have participated in trials around the world (Walters, 1999). Many scientists maintain that gene therapy has a great potential in the long term for the treatment of human diseases. But presently, clinically effective gene therapy is the exception.

When we discuss the ethical aspects of gene therapy, it is sometimes important to make a distinction between ethical issues raised by:
- gene therapy research, and
- gene therapy.

In the former case, special problems of research ethics arise. In the latter, social aspects of the use of gene therapy are of central importance.

In an ethical discussion, it is also important to distinguish different kinds of gene therapy. In the 1980s LeRoy Walters suggested a classification that has become very commonly used (Walters, 1986). He made a distinction between the following kinds of genetic modification:
(1) somatic cell gene therapy,
(2) germline gene therapy,
(3) somatic cell genetic enhancement, and
(4) germline genetic enhancement.

Research on somatic cell gene therapy is presently the only kind that is carried out by scientists, and it is not particularly disputed. The other kinds of genetic modification, however, are all, as indicated above, very controversial.

The classification is now being challenged. The distinction between somatic cell gene therapy and germline gene therapy is not completely strict. Somatic cell gene therapy might sometimes have unintended effects on the germline. Moreover, the line between therapy and enhancement is even more fuzzy. Should disease-related enhancement, e.g. improvement of the immune system above normal, be classified as enhancement or as therapy?

Recently, a supplement to the classification has also been suggested. It has often been presupposed that genetic interventions are limited to the nuclear genome. Given recent advances in genetics, however, the mitochondrial genome and diseases must be considered as well. This adds a new dimension to both somatic gene therapy and germline gene therapy (Richter, Bacchetta, 1998, pp. 304-306).

It might be disputed whether all these distinctions are ethically relevant. But it is quite clear that they are useful to keep in mind in any debate on gene therapy and its ethical implications, at least in order to understand the debate as such.

Somatic Cell Gene Therapy Research: Ethically Problematic?

How one judges somatic gene therapy depends to some extent on which prototypical cases one starts from. If somatic therapy is used as a prototypical case, somatic gene therapy appears to provide just certain new methods, not something radically different. Somatic gene therapy is viewed like any other somatic therapy, although the treatment is targeting diseases at a more fundamental level than ordinary treatment.

If, on the other hand, the natural order in terms of an unmodified genome is used as a prototypical case, then somatic gene therapy raises qualitatively new ethical problems. Proponents of this view may argue that the genome is such a fundamental part of the human person that any genetic intervention will affect the essence of what it is to be human. Other treatments only affect the phenotypic expression of the genome. Somatic gene therapy affects the genotype itself.

I am critical of this latter view but its prevalence among the general public makes it necessary to discuss the ethical acceptability of somatic gene therapy in itself. Another reason is certain events during the last couple of years that have made somatic gene therapy research controversial in a way that it never was during the previous decade. I am thinking of the deaths of people participating in somatic gene therapy trials.

Dead Patients and Failures to Report Adverse Events

In September 1999, Jesse Gelsinger, an 18-year-old patient participating in an adenoviral vector gene transfer clinical protocol at the Pennsylvania University, died. His death was not a result of his underlying condition, but seemingly due to an immunological reaction as a direct result of administration of a gene transfer product. Since this kind of research started in 1990, this is presumed to be the first death directly related to the gene transfer itself.[3] Since the Gelsinger case, however, there has also occurred another death of a patient participating in a somatic gene therapy trial, this time at Tufts University. Several other failures to report adverse events have also been disclosed.[4]

All this has made somatic gene therapy controversial to an extent it never was before. Even people who have previously been very positive to somatic gene therapy have been forced to at least rethink the issue. It is vital that responsible geneticists take this new situation seriously.

Moral Imagination: Arguments in Favor of and Against Somatic Cell Gene Therapy

So, is somatic gene therapy ethically acceptable? Can responsible geneticists carry out somatic gene therapy research? At the very least, before they do such research, they should use their moral imagination and take into account different viewpoints and arguments. I shall present the most important arguments in favor of as well as against, although rather briefly. Some recur in a slightly different form in the discussion about germline gene therapy and I shall analyze and discuss them more extensively in that context.[5] In my analysis I shall not distinguish between research on and use of somatic gene therapy. I shall simply assume that the arguments would be the same. As a contribution to dialogue, I shall also briefly comment on each argument. In order to facilitate ethical reflection, the presentation will be rather systematic.

There are basically two kinds of ethical arguments in favor of somatic gene therapy. They are based on the principles of beneficence and autonomy, respectively. The argument based on the principle of beneficence stresses the medical benefits of somatic gene therapy. A central goal of health care is to fight disease. Somatic gene therapy is a new way of doing this. Moreover, it could be one important way of bridging the 'therapeutic gap'. In my opinion, this argument is completely acceptable.

Other arguments are based on the principle of respect for autonomy. One argument stresses the autonomy of the scientists. Scientists should be free to do whatever research they want including research on somatic gene therapy. Scientific freedom is an important value and restrictions of this freedom will have negative effects. Another argument focuses on patient autonomy. If patients or the parents of ill children, request somatic gene therapy it should be provided given that there are sufficient resources. Another argument stresses the respect for reproductive autonomy. If prospective parents request somatic gene therapy on their fetus, this should also be provided on the same condition.

---

[3] *Teen Dies Undergoing Experimental Gene Therapy*, Washington Post, September 29, 1999, p. A01. See also URL: www.nih.gov/about/director/07122000.htm.

[4] *Gene Researchers Admit Mistakes, Deny Liability*, Washington Post, February 15, 2000, p. A03.

[5] A recent overview of the arguments can be found in Holtug, 1997.

The argument from scientific autonomy is not acceptable in itself. I have already discussed the question of scientific freedom in general. There should be some ethical restrictions on the choice of subject in scientific research. This does not mean, however, that somatic gene therapy research should not be conducted. It only means that this has to be justified by other arguments than one referring to scientific freedom. The other arguments from autonomy, however, are quite acceptable.

If we turn to the arguments against somatic gene therapy, they can be classified into three categories. The first includes arguments from respect for integrity. Some argue that the genome is such a fundamental part of the human person that any genetic intervention will affect the essence of what it is to be human. Other therapies affect only the phenotypic expression of the genome. Somatic gene therapy affects the genotype itself. This means that somatic gene therapy violates the integrity of the genome and thereby the integrity of the person. Sometimes this kind of argument is articulated in terms of 'playing God'. Humans are doing what only God should be doing. Another version talks about violating 'natural order'. The natural state of the human genome should not be tampered with.

These arguments from integrity are all open to serious objections. One general problem is that they seem to presuppose certain life views that are not generally shared. If you do not share these life views, they hold little weight. This in turn makes them less appropriate as a basis for the professional policy of the geneticists. However, I will not deal with these arguments right now. I will discuss them extensively in the section about germline gene therapy. In my opinion, they are not tenable as arguments against somatic gene therapy research.

The second category of arguments against somatic gene therapy includes arguments from the principle of nonmaleficence. They stress the unknown risks of somatic gene therapy. The specific risks of somatic gene therapy are of at least two types. Some arguments focus on scientific uncertainty and clinical risks, other arguments on social risks. Somatic gene therapy research will by necessity be surrounded by great scientific uncertainty due to its experimental nature. Moreover, there might be certain clinical risks. The transferred gene might be abnormally expressed leading to toxicity. Viral vectors might lead to infections. Cancers might emerge. These arguments should be taken very seriously as illustrated by the death of Jesse Gelsinger. It is clear that no scientist should carry out somatic gene therapy research without sufficient safety precautions. However, if the clinical risks can be expected to be minor, somatic gene therapy should be allowed. This is something that the principle of beneficence demands.

Other objections focus on social risks and slippery slopes. Somatic gene therapy in analogy with genetic testing and screening might lead to discrimination and stigmatization of persons who continue to live with diseases or disabilities that are subject to intervention with somatic gene therapy. Their human value might be reduced by such intervention, or at least they might feel so, which is bad enough. In this way somatic gene therapy might reinforce existent social discrimination of disabled persons. Many organizations for deaf and blind people are for this reason critical of somatic gene therapy. They are satisfied with their lives and do not want somatic gene therapy for themselves or their children (Lebacqz, 1999). Another social risk is that somatic gene therapy will be available only to rich. There are also warnings against a slippery slope to germline gene therapy and genetic enhancement. Moreover, there is the argument that somatic gene therapy reinforces 'geneticization'. Too much focus on genes might lead to neglect of the social causes of ill health.

The social risk arguments against somatic gene therapy are the most important. They all have a valid point, although they are not sufficient to reject somatic gene therapy. The risks of discrimination, stigmatization and reinforcement of social injustice are real. Somatic gene therapy should be accepted only if the prospects of counteracting such bad consequences are promising. The key question is if they are. It should be noted that these risks are not unique to somatic gene therapy. They are obvious also in health care in general. However, in health care in general they are not considered reason enough not to let those who are prepared to pay for such services do so. It could reasonably be argued that the same should hold for somatic gene therapy. On the other hand, we should work for making health care in general, as well as somatic gene therapy, available to as many 'ordinary' persons as possible. However, these social risks are not reason enough to not allow somatic gene therapy as such. Moreover, the critical views of certain disabled people should be respected. This is best achieved by stressing the importance of autonomy. Deaf and blind people, whether adults or children, should not feel any pressure to use somatic gene therapy. The government has the responsibility to counteract such pressure.

The slippery slope arguments should also be taken seriously, although the metaphor is not very good. A better metaphor would be a staircase–it opens up the possibility of stopping in time. Whether germline gene therapy and genetic enhancement should be accepted will be discussed later. What is decisive is that it is possible to effectively say 'no'. Finally, geneticization is always a risk. However, it can be and should be avoided. The risk for geneticization is not a decisive argument against somatic gene therapy. It is possible to combine an acceptance of somatic gene therapy with an awareness of the social causes of disease.

The third category includes, among other arguments, an objection that draws attention to the allocation of resources in biomedical research and health care. Somatic gene therapy research is very expensive and will probably be so for a long time. Somatic gene therapy will not for a foreseeable future be cost effective enough. It would be better if the resources are allocated to other fields, for instance genetically tailor-made drugs. This is also an important argument. But it does not seem decisive enough. One has to start somewhere and in all modern R&D the initial costs are very high. However, if somatic gene therapy holds a great potential, it could be well worth developing. In the end, somatic gene therapy could be very useful clinically.

Another objection stresses the limited success of somatic gene therapy research to date. The argument can be used by those who are completely against gene therapy research. But it can also be used as a way of supporting germline gene therapy research as a more promising alternative. I will come back to this issue below.

In conclusion, no decisive argument against somatic gene therapy has been found. Moreover, medical benefits and respect for patient autonomy outweigh the social risks. The social risks should, however, be taken seriously. Responsible geneticists should use their moral imagination to find ways of counteracting stigmatization, discrimination, reinforced inequalities, geneticization, and certain further developments that some label 'slippery slopes'. The question remains, however, how they should handle the fact of limited success in somatic gene therapy research.

'Ordinary' Problems of Research Ethics

If one accepts somatic gene therapy research, many traditional questions in research ethics become vital. Examples are: what diseases should be the prime targets for somatic cell gene therapy research? What risks and benefits are connected with experiments in somatic gene therapy? How should patients be selected in a fair manner? How can a voluntary, informed consent be secured? How should the privacy of the human subjects and confidentiality of medical information be protected?[6] This set of questions concerns research on and development of gene therapies. There is also another set of questions concerning gene therapy in a wider context. To what extent should society fund somatic gene therapy research? How can we see to it that not only rich people in the Western world get the opportunity to use this kind of therapy? Responsible geneticists have a duty to consider both sets of questions and try to find balanced answers.

Choice of Diseases and Selection of Patients

The first issue concerns priority-setting. Which diseases should be prioritized as subjects of research? Here, geneticists have to be aware of the fact that scientific and ethical aspects are intimately intertwined.

There are several possible criteria. One is severity. It would seem natural to choose more serious diseases before the less serious. However, things are not as simple. This becomes obvious when we turn to the next possible criterion, namely prevalence. Perhaps we should choose more common diseases before more severe? Some of the most severe diseases are very rare. It might be more ethical to put our efforts toward more common diseases. On the other hand, more rare diseases, whether more or less serious, might have better prospects of success, and this is the

[6] Cf. the NIH "Points to consider"; see URL: www4.od.nih.gov/oba/rac/guidelines/appendix_m.htm.

third possible criterion. If we prioritize disease with better prospects of success but are more rare, we might learn things that are essential for finding therapies to diseases with fewer prospects of success but are more common or more serious. However, there is also another possible criterion, namely lack of alternative treatments. It could be argued that we should prioritize such diseases for which there are no alternative therapies. On the other hand, it might be better to put our efforts toward some particular diseases despite the fact that there already are some therapies available. These diseases might have better prospects of success, be more common, or more be more serious.

These brief reflections show that the issue of priority-setting is very difficult. The easy case is when a disease fulfills all these four criteria. It is serious and common, it has good prospects of success, and there are no alternative treatments available. One suggestion is that we should look out for such diseases as prime targets for somatic cell gene therapy research. In the absence of such diseases, we should prioritize diseases with good prospects of success. Gene therapy research is very difficult and experimental. It would be wise to start with diseases to which therapeutic solutions can be anticipated to be more easily found than others.

Even when the geneticists have found a suitable disease as target for gene therapy research, there remains the problem of making a fair selection of patients who have that particular disease. Who should be recruited? With rare diseases, the selection is rather unproblematic. With more common diseases, fair selection can be very difficult.

Dialogue with Patients

As soon as they have made a fair selection of patients, the researchers face the problem of dialogue. They have to obtain informed consent. They have to inform about the nature of the experiment as well as of the risks and benefits of participating. This is common to all research involving human subjects. However, in gene therapy research special problems arise. The relative lack of success in gene therapy research to date makes it important not to overstate the possible benefits to the patients. Moreover, the patients must be informed about risks in terms of media attention. When it comes to participation in research of this experimental nature, breakthroughs may create headlines in media. This might raise difficult problems concerning the privacy of the human subjects and the confidentiality of the medical information. The patients must be informed about this risk in the process of consent. Moreover, certain measures must be taken in order to prevent such problems.

Special problems are raised by the possible participation of children in gene therapy research. One could maintain the view that only adults should participate in this kind of research until its success and safety are well established. However, certain genetic diseases are expressed in children very early in life and it would be unreasonable to ignore their need for treatment. But, of course, the parents should give their vicarious permission and, if possible, the assent of the children should be obtained. With 'assent' I mean the not legally effective authorization of the child. There might be certain risks with the experimental treatment, but as long as these are only minor, they should not be viewed as decisive enough to halt the participation of children. Moreover, even if the risks are slightly higher than minor, it might be justified to let the children participate, particularly if their clinical condition is life-threatening (cf. Levine, 1996).

The Therapeutic Gap

Let me also make a comment on the 'therapeutic gap', i.e., the gap between our immensely increased ability to get information about genetic diseases through genetic testing and our ability to give efficient treatment. There are two ways of responding to this gap. First, we might avoid genetic testing for diseases for which there is no cure. This is the way preferred by many people when asked in genetic counseling whether they want the information. For the future, however, this is, of course, unsatisfying. Therefore, and secondly, we need to do more research on gene therapy and other therapies, for example on genetically tailor-made drugs. In fact, the therapeutic gap is among the best reasons for carrying out such research.

Germline Side Effects

Somatic gene therapy might in some cases have germline side effects. Such side effects are not new in medical treatment. Radiation in cancer treatments is one example. This raises two kinds of ethical questions. The first one is whether this constitutes a decisive argument against all somatic cell gene therapies. This is not the case. If germline gene therapy is ethically acceptable in principle, then germline side effects of somatic gene therapy do not in themselves constitute a decisive argument against somatic gene therapy. If germline gene therapy is not acceptable, then germline side effects of somatic gene therapy might nevertheless be acceptable under some circumstances. The benefits of somatic gene therapy might outweigh the costs.

The second question is whether and, if so, how to avoid such side effects. If one is critical of germline gene therapy, then one should try to avoid them. However, it is possible to accept germline gene therapy and nevertheless be critical of germline side effects of somatic gene therapy. One might simply not want such side effects in certain particular cases. So, from both perspectives, the question is how to avoid the side effects. One way is to abstain from carrying out the somatic gene therapy; another is to improve the technique of the gene therapy.

Criticisms of 'Overselling' and Limited Success

Many geneticists are convinced that gene therapy has an extraordinary potential in the long term for the treatment of human diseases. However, in December 1995 a Panel appointed by the National Institutes of Health (NIH) to make an assessment of the research on somatic gene therapy within the US concluded that clinical efficacy had not yet been definitely demonstrated in any gene therapy protocol. Moreover, the Panel also found that the scientists themselves had contributed to the misconception that gene therapy had been much more successful than it actually had been. Another conclusion of the Panel was that much research is left to do, not least purely basic research on the pathophysiology of diseases. It also recommended that the scientists be more restrained in the 'selling' of their research.[7]

The 'overselling' of research raises, of course, questions of moral responsibility of the gene therapy researchers. To oversell is to deviate from truth and to do this in the promotion of research is not morally acceptable. However, the overly optimistic attitude of certain geneticists can easily be understood. The relative success in the genetic engineering of microorganisms, plants and animals might be one reason. The overselling is also understandable as an effort of raising money to promising research. On the other hand, it is not excusable. It is a matter of misleading funding agencies, other scientists, politicians and the public.

One of the objections to somatic gene therapy research mentioned above focuses on its limited success. This needs an explanation. Why has the success of this kind of research been so low? One explanation that has been suggested is that somatic gene therapy involves more practical and technical problems than was originally thought. The optimism at the beginning depended on a lack of understanding of the difficulties of applying results from research on animals to research on humans. Once the practical and technical problems are solved somatic gene therapy will make fast progress (Smith, 1999; Brenner, 1999). Another explanation is that there are some problems in basic science concerning the pathology and physiology of diseases that have to be solved. Once these problems are solved somatic gene therapy will eventually be successful. This was the view stressed by the NIH Panel.[8] Thirdly, we have the explanation that the human genome is too complex to be manipulated other than in a few special cases (Botstein, 1997). Related to this is the fourth explanation, proposed by Richard Strohman and others, that somatic gene therapy has been based on certain fundamental assumptions that are simply false. According to the 'central dogma' of genetics, DNA causes RNA which causes proteins. Things are actually much more complex. The central dogma has to be revised. However, once this is done it will become obvious that somatic gene therapy only to a very limited extent will be successful (Strohman, 1993; Graumann, 1999).

All these explanations have their valid points. The first two accurately stress the lack of knowledge, scientifically and technically. But this is no reason to stop doing

[7] See URL: www4.od.nih.gov/oba/rac/panelrep.htm.
[8] See URL: www4.od.nih.gov/oba/rac/panelrep.htm.

somatic gene therapy research. On the contrary, the scientists should continue even harder. If the obstacles are eventually overcome, the medical benefits will be vast. The last two explanations rightly stress the vast complexity of the subject. However, the extent to which somatic gene therapy will be successful cannot be judged a priori. It can only be discovered in practice. Proceeding is the only way of discovering exactly how complex it all is and where the problems actually lie. Moreover, the success in animal models suggests that the research will eventually be successful, and encouragingly enough, a clinical breakthrough has recently been reported.

A Recent Clinical Breakthrough

A promising gene transfer has been presented for the treatment of infants with one form of X-linked severe combined immune deficiency (SCID). The infants appeared to have their disorder corrected by the transplantation of their own bone marrow cells, into which a normal copy of the disease-causing gene had been introduced by retroviral gene transfer. They developed sustained immunological function similar to that seen in SCID patients transplanted with normal allogeneic bone marrow cells. The infants, who were otherwise forced to live inside protective 'bubbles', are healthy and living normal lives several months after being treated with this somatic gene therapy (Cavazzana-Calvo et al., 2000).[9]

This achievement has been characterized by some as the clinical breakthrough for which so many researchers have been waiting for so long.

Dead Patients–Cured Patients: A New Situation

The reports about dead patients in combination with the clinical breakthrough have created a new situation in gene therapy research. Several ethical and policy problems arise. A key issue is: how should the scientific community handle this complex situation with scandals as well as success?

With regard to the scandals, the short-term problem is: what does it imply to take the deaths seriously? What should be the professional policy? Should gene therapy researchers in general interrupt their research or just be more cautious? The latter seems enough, but there also seems to exist a need for more dialogue within the scientific community about the tendency of taking risks for commercial or professional reasons. A more general long-term problem is whether stricter governmental oversight is the right approach or whether the scientific community itself should assume responsibility. The latter alternative seems unavoidable, if the scientific community wants to retain credibility. Stricter governmental oversight

---

[9] See also *Genetic Therapy Apparently Cures 2*, Washington Post, April 28, 2000, p. A01. See also URL: www.nih.gov/about/director/07122000.htm.

might be necessary in some countries, but what is truly important is that the worldwide scientific community find ways to monitor the research and open a serious dialogue with the public about risks and benefits.

With regard to the clinical breakthrough, it is clear that researchers all over the world should be encouraged. But it is also important to remain patient and not be overly optimistic. The complexity of the subject matter is vast. Success in one area is no guarantee for success in another. The criticism of the NIH Panel in 1995 against 'overselling' the research should still be taken seriously.

2. GERMLINE GENE THERAPY

A few years ago germline gene therapy was not truly discussed other than very hypothetically. In the last years, however, the issue has been more openly discussed (de Wachter, 1993, p. 174). But even those who would be willing to accept it 'in principle' often state that it would not be clinically justified since embryo selection based on preimplantation genetic diagnosis would be a better alternative in most cases. However, several very well-known geneticists and gene therapy researchers have recently expressed the view that maybe germline gene therapy research could be much more successful than somatic gene therapy. At a symposium in Los Angeles on March 20, 1998, W. French Anderson said: "The unfortunate fact is that, with the exception of a few anecdotal cases, there is no evidence of a gene therapy protocol that helps in any disease situation." He said this not in order to discourage further research. On the contrary, he takes this as an incentive to develop germline gene therapy. As James Watson said at the same symposium, "[o]ne thing that seems pretty obvious is that germline therapy will probably be much more successful than somatic. If we wait for the success of somatic we'll wait until the sun burns out."[10] It should be noted that this was stated before the clinical breakthrough mentioned above. Nevertheless, this interest in germline gene therapy is important. In combination with the moratorium called for by the AAAS report mentioned at the beginning of the chapter, it brings, to be sure, this kind of gene therapy to the forefront of the ethical discussion. In addition, the possibility of germline gene therapy on humans has recently come one step closer with a successful germline gene transfer into a rhesus monkey. Such genetic modifications have been carried out on mice and rats for several years, but they have now been achieved on an animal genetically even closer to humans (Chan et al., 2001, pp. 309-312).

Dialogue and Imagination

---

[10] See URL: www.ess.ucla.edu:80/huge/report.html.

It is vital that responsible geneticists take the issue of germline gene therapy seriously, take part in a dialogue with the general public, and develop a well-considered professional policy. Reflection by responsible geneticists on this issue requires, however, more effort in using scientific and moral imagination than most other ethical issues raised by modern genetics. Scientific imagination is needed, because germline gene therapy has not as yet been carried out on humans. It is still only a possibility for the future. Moral imagination is necessary, because this kind of modification is in a sense more radical than all other medical interventions. It is a modification of the genome that is inherited by future generations.

Official Policy Regarding Germline Gene Therapy

In many countries, germline gene therapy research is legally prohibited. Moreover, many official policy documents do no allow for germline gene therapy. For instance, in the 1996 *Convention of Human Rights and Biomedicine* issued by the Council of Europe, germline gene therapy is by implication not allowed. According to Article 13, genetic intervention is allowed (Council of Europe, 1996):

> ... only if its aim is not to introduce any modifications in the genome of any descendents.

On the other hand, some official policy documents are more liberal. In UNESCO's *Universal Declaration on the Human Genome and Human Rights* (1997), it is stated that:[11]

> [r]esearch, treatment or diagnosis affecting an individual's genome shall be undertaken only after rigorous and prior assessment of the potential risks and benefits pertaining thereto and in accordance with any other requirement of national law (Article 5).

There is no prohibition against germline gene therapy. The issue is left to legal regulation in different countries. This means that the Declaration sees nothing objectionable in principle with germline gene therapy. To be sure, the *Declaration* states that:[12]

> [p]ractices which are contrary to human dignity, such as reproductive cloning of human beings, shall not be permitted (Article 11).

But no reference is explicitly made to germline gene therapy as a practice that is contrary to human dignity. It only states that germline intervention *could* be contrary to human dignity: [13]

> The International Bioethics Committee of UNESCO should... give advice concerning the follow-up of this Declaration, in particular regarding the identification of practices that could be contrary to human dignity, such as germ-line interventions (Article 24).

---

[11] See URL: www.unesco.org/ethics/.
[12] See URL: www.unesco.org/ethics/.
[13] See URL: www.unesco.org/ethics/.

Furthermore, in the *Declaration of Inuyama*, germline gene therapy is explicitly accepted. It comments on germline gene therapy as follows (CIOMS, 1995):

> The modification of human germ cells for therapeutic or preventive purposes would be technically much more difficult than of somatic cells and is not at present in prospect. Such therapy might, however, be the only means of treating certain conditions, so continued discussion of both its technical and its ethical aspects is essential. Before germ-line therapy is undertaken, its safety must be very well established, for changes in germ cells would affect the descendents of patients.

The opinions of bioethicists and geneticists are also divided. As indicated above, earlier most of them were negative but now a shift seems under way. Let us now analyze the main arguments in favor of and against germline gene therapy. I shall start by characterizing the debate.

Character of the Dispute

Responsible geneticists should use their moral imagination to consider different arguments in favor of and against germline gene therapy.[14] Many of the arguments can be used in a more or less radical way. They can be used in favor of or against some or all kinds of gene therapy. This fact makes them hard to assess. Many of them are extensions of commonly recognized ethical principles such as respect for integrity and nonmaleficence. From the point of view of imaginative casuistry, the key issue is how far and in what direction these metaphors can reasonably be extended.

From this perspective, it is also obvious that the problem of germline gene therapy can be framed in different ways. This framing might affect the discussion. It might make a difference whether the issue is framed in terms of an abstract issue of academic philosophy, scientists searching for new knowledge, health care officials aiming at cost effective health care, the government developing an official policy, or prospective parents wishing their children the best start in life. Below, I shall focus on the professional policy of the geneticists. To be sure, this book is about the moral responsibility of this group. The comments and proposals are meant as contributions to an interdisciplinary dialogue.

Moral Imagination: Arguments in Favor of Germline Gene Therapy

---

[14] For a recent overview of the arguments, see Holtug, 1997. See also Walters, Palmer, 1997, pp. 80-88, and Mieth, 1999.

As was the case in my analysis of arguments related to somatic gene therapy, the presentation will be rather systematic. The reason is to facilitate ethical reflection. In my reconstruction, there are three main types of argument in favor of germline therapy, and, as in the discussion of somatic gene therapy, I make no distinction here between research on and use of the therapy.

The first kind starts from the principle of beneficence and stresses that germline gene therapy has many medical benefits. The second category takes the principle of autonomy as a point of departure. One argument stresses the autonomy of the scientists, another the autonomy of prospective parents. There is also a third category including other kinds of arguments. Throughout this discussion it is presumed that the germline gene therapy is sufficiently safe. This may, of course, be problematic. Some would say that the problem is precisely that we do not know for certain whether it can be safe.

Arguments from Beneficence

Many arguments focus on medical benefits. One argument stresses the medical necessity of germline gene therapy. As we saw above, the *Declaration of Inuyama* argues that (CIOMS, 1995):

> [s]uch therapy might...be the only means of treating certain conditions.

Examples could be cystic fibrosis, which affects many different organs and disparate cell types, and the Lesch-Nyhan syndrome, which is expressed in the nonremovable and nondividing cells of the nervous system (Walters, Palmer, 1997, pp. 62-63). In cases like these, genetic intervention on the totipotent embryo is the only means of achieving successful treatment, because it affects all cells.

Another argument points out that germline gene therapy can provide true cure for many genetic diseases. Other therapies just treat symptoms of diseases or are merely palliative. By genetic intervention on the germline there will be no phenotypic expression whatsoever, because well-functioning genes will be added (in cases of recessive disorders) or the disease genes replaced by properly functioning ones (in dominant disorders). This must be the ultimate goal of medicine. In addition, it is also more in line with the general goal of health care of curing diseases than terminating the life of fetuses or selective discard of preimplantation zygotes.

Other arguments focus on the preventive efficiency of germline gene therapy. Germline gene therapy would be a way of preventing the transmission of recessive disorders to the next generation. It offers parents who are both afflicted by recessive genetic disorders or are both carriers of recessive genetic disorders the possibility of preventing transmission of these to their children. Moreover, germline gene therapy would be more efficient than repeating somatic gene therapy (or other therapies) generation after generation. By germline gene therapy it would be possible once and for all to eliminate a genetic disease in a family line. A third argument expresses explicitly a kind of eugenics. It stresses that germline gene therapy makes possible a reduction of the incidence of certain genetic disorders in the population.

How should these arguments be assessed? The first is in itself acceptable. It is quite in line with the therapeutic imperative. Germline gene therapy might be a medical necessity, i.e. the only way of treating certain genetic diseases. However, it would probably be used only very rarely. In most cases preimplantation selection would be a better alternative. The second argument is also acceptable. True cure is to prefer above mere symptomatic and palliative intervention. In both these arguments, germline gene therapy is used for phenotypic prevention.

However, germline gene therapy may also be used for genotypic prevention as can be seen in the argument that germline gene therapy is a way of preventing the transmission of recessive disorders to the next generation, and the argument that germline gene therapy will make it possible to once and for all eliminate a genetic disease in a family line. In Chapter 6 I will return to this issue and argue that genotypic prevention is acceptable only on the condition that it is requested by prospective parents. Mandatory germline gene therapy and even the active recommendation of such therapy by the geneticists or the government is unacceptable. The reason for this is that these measures would violate the reproductive autonomy of the prospective parents. Moreover, this means that the argument that germline gene therapy makes possible a reduction of the incidence of certain genetic disorders in the population cannot be accepted. Using germline gene therapy as a public health measure violates the reproductive autonomy of the prospective parents. It implies a social program that puts prospective parents under certain pressure, even if it is voluntary.

Arguments from Autonomy

The second type of argument, which we in fact have already met in our brief discussion of somatic gene therapy, is based on the principle of respect for autonomy. One argument stresses the autonomy of the scientists. Scientists should have the freedom to do whatever research they want including germline gene therapy research.

Another argument stresses the respect for reproductive autonomy. Prospective parents should have the opportunity to choose whatever reproductive technology they want. Consequently, they may request germline gene therapy and this should be accepted at least in a private health care system in which they pay for themselves. Requests in a public health care system could perhaps be turned down with reference to limited resources. Examples of situations in which they might request germline gene therapy have been given above. In addition, they might want to avoid selective abortion or preimplantation selection, or avoid producing children who are carriers of genetic disease.

As I have already stressed earlier, the first argument is not acceptable in itself. There should be some ethical restrictions on the choice of subject in scientific research. But this does not mean that germline gene therapy research should not be carried out, merely that it has to be justified by other arguments.

The second argument, however, should be taken seriously. Reproductive autonomy is extremely important. But there has to be certain restrictions. These restrictions should be disease-related. Germline interventions on nondisease characteristics should not be accepted (see pp. 248-252, 255-257).

Other Kinds of Argument

There are also other kinds of argument in favor of germline gene therapy. One stresses the impossibility of stopping the development of germline gene therapy (Silver, 1997, pp. 10-11). Parental requests of such therapy combined with commercialism have a dynamic of their own which makes it virtually impossible to stop. The important thing is to adjust to this development and make it as smooth as possible.

This argument is too deterministic. I have expressed the view that decisions made by individuals and small groups can affect the course of technological development, at least to some extent (see pp. 55-58). But, of course, this 'technological voluntarism' has certain limits. Our expectations have to be realistic. Parental requests and commercialism are strong forces in technological development. The extent to which we can affect future developments can only be discovered in practice.

According to another argument in favor of germline gene therapy, the prospect of success seems to be much greater in germline gene therapy research than in somatic gene therapy research. At least this was the opinion of some important geneticists and gene therapy researchers at the above mentioned conference in Los Angeles in 1998. Because of this prospect it is argued that responsibility in health care demands that germline gene therapy research is started.

I do not subscribe to the pessimistic view of somatic gene therapy. But if the optimistic view of the prospect of germline gene therapy has any substance, this argument may be important. But, of course, the argument is not decisive in itself. The mere prospect of success of the technology does not suffice. The argument has to be combined with other arguments about the benefits and risks.

Moral Imagination: Arguments Against Germline Gene Therapy

Now, we turn to arguments against germline gene therapy. I have already briefly presented some of them in the context of somatic gene therapy. I shall now discuss them more extensively and focus exclusively on germline gene therapy.

I find it useful to classify the arguments against germline gene therapy in three categories. The first includes arguments from respect for integrity. One example is the argument that every human being has a right to an untampered genome. The second category includes arguments from the principle of nonmaleficence. These arguments stress that certain kinds of harms, for example biological harms or social harms, can be anticipated as consequences of germline gene therapy. I call these arguments 'risk-oriented' arguments. The third category includes other kinds of arguments referring to consent of future generations, fair allocation of resources, and alternative strategies.

Arguments from Respect for Integrity

We have found that integrity is a metaphor that has been extended in many different ways. The term has its root in a Latin word meaning i.a. 'wholeness' and 'untouchedness'. 'Integrity' has, thus, the connotation of 'untouched wholeness'. The principle of respect for integrity has been applied to many different kinds of 'wholes', e.g. the Creation, Nature, persons, and genomes. Each of these has been considered a 'whole' that should not be 'touched', broken up, or violated (see pp. 47-49). Arguments from respect for integrity play a central role in certain official policy documents on germline gene therapy and in certain theological statements. The genetic patrimony has an integrity that should be respected. Germline gene therapy is an infringement on this heritage. Let me reconstruct some arguments belonging to this category.

One argument against germline gene therapy has a theological character and views such intervention as a violation of divine order. The human genome is interpreted as God's creation, and respect for the integrity of this creation requires that no germline gene therapy should be carried out. There are limits to what humans are allowed to do. Sometimes this argument is presented in terms of 'playing God' in the sense that humans are doing what only God should do.

There is also a secular version of this argument. Here germline gene therapy is viewed as a violation of natural order. In doing this kind of therapy humans are doing what humans should not do. The natural state of the human genome should not be tampered with. The integrity of the human genome in its natural state should be respected. According to a specific version of this argument, minor changes in the gene pool might accumulate over the generations with the result that the population sometime in the future is no longer human. Leon Kass, a well-known proponent of a more 'natural science' expresses this idea as follows (Kass, 1972, p. 61):[15]

---

[15] Quoted in Munson, Davis, 1992, pp. 149.

> It may... mark the end of *human* life as we and all other humans have known it... We are ourselves human beings; therefore, we have a proprietary interest in our survival, and our survival as *human beings*.

Sometimes there is also a reference to the wisdom of evolution. The human genome is the result of millions of years of evolution. It should therefore not be tampered with (President's Commission, 1982, pp. 62-63).

These arguments are all open to serious objections. The main objection is that the principles to which they appeal are ambiguous. These principles can be extended in many different ways. In fact, they can be used in arguments in favor of as well as against germline gene therapy. This holds for respect for divine order, respect for natural order as well as respect for the wisdom of evolution. Whether they are used in favor of or against germline gene therapy depends on other moral principles that must be imported to the discussion. Moreover, some of them are based on certain life views that are not generally accepted. If these life views are not accepted, the arguments carry little weight. This means that they are no good basis for the professional policy of the genetics community. In addition, there are also special problems with some of them.

The argument from respect for divine order depends on whether there is a divine order, and, if so, on how we conceive the divine order more precisely. When the argument is presented in terms of 'playing God', and this expression is used quite literally, it is presupposed that the divine order, as it is before germline gene modification is carried out, should remain as it is. However, it is possible to argue from respect for divine order that germline gene therapy is morally acceptable and perhaps even an obligation. The reason for this is that according to some theologians humans should be co-creators with God. To co-create could imply fighting disease including genetic disease, and a means of fighting certain genetic diseases could be to do germline gene therapy. This means that the argument from respect for divine order is not sufficient in itself. The mere reference to respect for divine order is not enough. It has to be shown that respect for divine order excludes germline gene therapy. But this requires other arguments. Another objection starts from the fact that this argument can be used against somatic cell gene therapy as well. If someone accepts somatic gene therapy but not germline gene therapy with reference to the 'playing God' argument, this seems incoherent. In order to avoid this accusation, the proponent has to show that there is a morally relevant difference between somatic and germline gene therapy, but this requires yet another argument.

The argument from respect for natural order is open to a similar objection. It can be used to argue in favor of germline gene therapy as well as against it. In the former case, it is assumed that the natural order is the same as a well-functioning order and that such an order excludes genetic disease. Therefore, in order to keep the well-functioning order germline gene therapy might be necessary. This means that mere reference to respect for natural order is not sufficient. It has to be shown that respect for natural order is incompatible with germline gene therapy. This, however, requires other arguments. Another objection to the argument from respect for natural order focuses on the implicit distinction between natural and unnatural intervention. According to one interpretation, all human interventions are natural in the sense that they are part of the natural order. In this case, no human interventions are unnatural. But then the argument does not work as an argument against germline gene therapy. According to other interpretations, there might be a difference between natural and unnatural interventions but in these cases it has to be shown that germline gene therapy is unnatural. This requires, however, an independent argument. The mere reference to respect for natural order is not enough. A third objection stresses that the argument from respect for natural order rests on the assumption that natural order is good. However, in order to avoid the 'naturalistic fallacy' it has to be shown by an independent argument that this is the case. The argument is not sufficient in itself.

The argument from the wisdom of evolution is also unacceptable. Even if there is some 'wisdom' in natural selection in general, and even if there is some 'wisdom' in that natural selection has made us vulnerable to genetic diseases, we cannot trust this 'wisdom' as a guide for our actions. As a matter of fact, many other medical practices have actually counteracted evolution. People who otherwise would not have been favored by natural selection are now able to reproduce and propagating their disease genes. And most of us find this completely acceptable. It is the humane thing to do.

Other arguments focus on human dignity and human rights. These arguments– and others–are implicitly found in the *Convention on Human Rights and Biomedicine* issued by the Council of Europe in 1996. The general foundation of the *Convention* is presented in Article 1 (Council of Europe, 1996):

> Parties to this Convention shall protect the dignity and identity of all human beings and guarantee everyone, without discrimination, respect for their integrity and other rights and fundamental freedoms with regard to the application of biology and medicine.

These principles are in the subsequent articles applied to different issues in biomedicine. In Article 13 there appears an implicit disclaiming of germline gene therapy. The concept of dignity and the concept of rights have different metaphorical connotations that affect how they are used in arguments against germline gene therapy. The metaphor of dignity suggests that germline gene therapy does not respect persons as ends in themselves. The financial metaphor of rights implies that each person, including also future persons, has the credit of being born with an untampered genome. Germline gene therapy ignores this credit. This can be compared to the metaphor of integrity that is also mentioned in Article 1. This metaphor suggests that the genome is a whole that should be left untouched. Germline gene therapy is wrong because it breaks up this whole.

The concepts of human dignity, integrity and rights are often considered to be intimately related. They often come in a package. Many would maintain that if you have one of them, you have the others. For example, it is sometimes stated that human rights are based on the concept of integrity. This kind of reasoning is seen in a report to the Council of Europe from 1982, which I have already quoted above (Council of Europe, 1982):

> The principle on which all existing human rights are based is the idea of man's integrity... A person's chromosomes, genes, etc., must clearly be a part of what is covered by this integrity just like his spirit, flesh, and blood.

Here we clearly see that the genome is part of an individual's integrity and should be respected.

If we turn to the UNESCO *Declaration*, we have already seen above that it refers to human dignity in relation to germline gene therapy. But the Declaration only indicates that germline gene therapy *could* be contrary to human dignity (Article 24); it does not argue that it truly does.[16]

Finally, there is an argument against germline gene therapy that refers to human genes as 'a common human heritage' like sea-beds or famous pieces of art. This has been stressed by Emmanuel Agius in the following manner (Agius, 1990, p. 140):

> Human genes are common to all generations... They are a common heritage because they are handed down from on generation to another... No generation has therefore an exclusive right of using germline therapy to alter the genetic constitution of the human species.

The idea of a common human heritage is central also in some official documents. Article 1 in the UNESCO *Declaration* states:[17]

> The human genome underlies the fundamental unity of all members of the human family, as well as the recognition of their inherent dignity and diversity. In a symbolic sense, it is the heritage of humanity.

We see that the principle of the human genome as a human heritage is fundamental to this *Declaration*. However, two things should be noted. First, the human genome is the heritage of humanity only "in a symbolic sense." This is, however, quite in line with my analysis of all these concepts. From the perspective of cognitive semantics, it is no surprise that they are all metaphorical. Second, this principle is in this *Declaration* not used as an argument against germline gene therapy, but only as justification of a general protection of the human genome. No explicit objection to germline gene therapy is made.

The arguments from human dignity, human rights, and a common human heritage are all unsatisfying. My main reason is the same as the one put forward against the arguments from respect for divine order and respect for natural order. The principles to which they appeal are all ambiguous and can be extended in many

---

16 See URL: www.unesco.org/ethics/.

17 See URL: www.unesco.org/ethics/.

different ways. In fact, they can be appealed to in arguments in favor of as well as against germline gene therapy. This holds for respect for human dignity, respect for human rights, as well as respect for the human genome as a common heritage. Whether they are used in favor of or against germline gene therapy depends on which prototype case is chosen as a point of departure and on other arguments that are imported to the discussion. In addition, there are also special problems with some of the arguments.

The principle of human dignity can be appealed to both by those against and those in favor of germline gene therapy. In the use described above, germline gene therapy does not respect persons as ends in themselves. However, it is quite possible to say that human dignity is respected precisely through this kind of therapy. But this is true only when this kind of treatment has the goal of helping a preembryo with a disease that can only be treated this way. If the preembryo is used only as a means for helping future generations, then this is contrary to the principle of human dignity. This means that the argument from human dignity is acceptable only as an argument against certain kinds of germline gene therapy, not all of them. However, the principle gives us some guidance concerning good and bad uses of germline gene therapy. We should not use germline gene therapy only as a means of helping future generations. It should always aim at also helping the preembryo (embryo, fetus, child) itself.

I also find the argument from human rights unsatisfying. If human beings truly have a right to an untampered genome, this right might conflict with other rights, e.g. the right to treatment. As stated above, germline gene therapy might be the only way of treating certain diseases. It is not at all self-evident that the right to an untampered genome outweighs the right to treatment. Moreover, an appeal to the right to treatment can be used not only on behalf of the preembryo at hand but also on behalf of future generations. This means that the argument from a right to an untampered genome has to be supplemented by another argument showing that this right outweighs the right to treatment of the preembryo as well as of future generations. In my opinion, we have reason to maintain that the right to treatment is more important. However, as I shall argue below, we have also a reason to make a distinction between germline gene therapy aiming at helping a preembryo and germline gene therapy aiming at helping future generations.

The argument from a common human heritage is also untenable. Eric Juengst has presented the following objections, all of which I find well-founded (Juengst, 1998b). A major problem is that it is unclear what exactly is thought to be a common human heritage. One possibility could be the human germline. However, from a biological perspective, there is no intergenerational human germline. Sexual relationships were substituted for the asexual germline relationships of dividing cells millions of years ago. Another possibility is the human genome. However, the human genome is an abstraction, like the human skeleton, not a natural object. Protection of an idea seems not to be what the proponents of this argument have in mind. A more reasonable interpretation is that they intend to refer to something actually existent in nature. This means that the human gene pool seems to be a better option. However, the gene pool constantly changes. Every reproductive choice 'tampers' with the gene pool available to the next generation. The problem is that this argument gives no guidance on how to distinguish between good and bad choices. Moreover, this means that the argument from a common human heritage, like the other arguments, can be used in favor of as well as against germline gene therapy. The idea of the human gene pool as a common human heritage can be used to justify germline gene therapies aiming at reducing the incidence of disease genes in the human gene pool. Such interventions could be viewed as a way of cultivating and taking care of the common heritage, a policy that the argument was originally designed to protect from. If such a use is not accepted by the proponents of the argument, they have to supplement it with other considerations or use another argument. An appeal to common heritage is not sufficient in itself.

Risk-Oriented Arguments

The second category of arguments against germline gene therapy is risk-oriented. These arguments stress the unknown risks of germline gene therapy. We live in what Ulrich Beck calls a 'risk society', i.e., a society with 'manufactured risks' (Beck, 1992). It is a society that creates risks through its technology and accepts some degree of risk in all its technological activities. According to this category of arguments, our high technological society is characterized by an inappropriate and irresponsible handling of risks. Germline gene therapy is a possibly dangerous part of this risk society. We should try to reduce the level of risk in society, rather than increasing it with such therapy. This view is based on the principle of nonmaleficence.

The specific risks of germline gene therapy are of at least three possible types. Some arguments focus on scientific uncertainty and clinical risks, others on long-term biological risks, yet others on social risks.

Germline gene therapy research will by necessity be surrounded with great scientific uncertainty due to its experimental nature. Moreover, there might be some clinical risks. As in the similar argument against somatic gene therapy presented above, the transferred gene might be abnormally expressed leading to toxicity, viral vectors might lead to infections, and cancers might emerge.

The argument from scientific uncertainty should be taken seriously but it is not decisive. No one proposes germline gene therapy research until sufficient safety precautions are met. Such research should only be permitted when the safety of somatic cell gene therapy is very well established. Moreover, it should only be allowed for diseases for which such therapy is the only or most efficient means of treatment. And, of course, clinical use of these forms of germline gene therapy should be permitted only when their safety can be guaranteed with high certainty.

The clinical risks should also be taken seriously. The research must proceed very cautiously. Germline gene therapy should be accepted only if the clinical risks are very limited. If they can be expected to be very limited and there are no other reasons against it, it should be allowed. It could help prospective parents and their preembryos. This is something that the principle of beneficence demands.

There are also warnings against long term biological risks. There might occur irreversible changes in the gene pool. In many cases, however, the changes resulting from germline gene therapy might be possible to reverse. But if the harmful consequences are discovered only after a considerable period of time when several future generations have been affected, this might not be possible. Moreover, gene therapy on the germline might lead to a reduced genetic variation, i.e., to a reduction of the gene pool. Such reduction might have bad consequences for the human species, because genes for certain diseases may at the same time give protection against other diseases. One example is the gene for sickle cell anemia that produces resistance to malaria. We are at risk of losing genes with possible fitness enhancing functions in future environments if we remove them with germline gene therapy. A third argument stresses that germline gene therapy will affect the well-being of future generations in a way that cannot be accurately predicted. We simply do not know the risks. Therefore, it would be irresponsible to carry out germline gene therapy.

The biological risks should be taken very seriously, but at the same time it must be emphasized that they should not be overestimated. We have at least some knowledge about how genes and evolution work; we should not overstate our ignorance. Moreover, future generations will probably be able to handle the modifications better than we can. Changes that might seem irreversible to us might not be to them. It might be true that germline gene therapy reduces genetic variation. But it is also possible that it will increase genetic variation, particularly if enhancement techniques are developed. It is also possible that the reduction of disease genes might lead to a reduction of protection genes. However, this is all speculation. Moreover, in cases like sickle-cell anemia and malaria, the problem is not difficult. If we learn how to prevent sickle-cell anemia with germline gene therapy, we should not worry about malaria. We already know how to treat malaria by nongenetic means. But we do not have the means of treating sickle-cell anemia. Finally, we should not magnify the risk of harming future generations. We should be very cautious, but not overly cautious. Scientists should be allowed to start germline gene therapy research only when we have no special reasons to fear particular risks, and only if we want to help patients at hand and think that germline gene therapy is the only way of doing this. This may imply, however, that it is never started at all.

Another cluster of arguments focuses on social risks and slippery slopes. According to one argument, germline gene therapy in analogy with genetic testing and screening might lead to discrimination and stigmatization of persons who continue to live with diseases or disabilities that are subject to intervention with germline gene therapy. Their human value might be reduced by such intervention, or at least they might feel so which is bad enough. Germline gene therapy might reinforce existent social discrimination of disabled persons.

Another social risk is that germline gene therapy will be available only to the rich. Being a very expensive technique germline gene therapy will only be an opportunity for a few people in the Western world. The distribution of this medical resource is at risk of being unfair, nationally as well as globally. Germline gene therapy would exacerbate social inequalities. This is not in line with the principle of justice.

There are also arguments warning against a slippery slope to enhancement. There is a risk that if we start allowing germline gene therapy we will after a time allow more and more, and eventually genetic enhancement on the germline will be accepted. Most parents would want to give their children as good a start in life as possible. This might lead some of them to request genetic enhancement. Such techniques might create children who are, for instance, more intelligent, physically stronger and more 'physically attractive' than they would otherwise be. Children with desirable traits can be ordered by couples, being 'designed' in much the same way as industrial products. As with germline gene therapy, such enhancement would probably for a foreseeable future be an opportunity only for the rich. We would get a 'new genetic upper class' born with far better potentials than other people. This is hardly compatible with the principle of justice and would probably also create too much tension in society. Even if our intentions from the very beginning are strict, it might be impossible to keep the regulations strict in the long run. The use of germline gene therapy might not be possible to control, at least not globally.

This argument is taken a step further by an argument that warns against possible misuse of genetic enhancement by dictators or totalitarian states. A common version of the argument speculates about what could have happened if Hitler had knowledge of genetic engineering techniques. By enhancing techniques he could have created elite soldiers whom no one could stop. Genetic enhancement is a dangerous instrument in the hands of a dictator.

Another important type of argument warns against a new eugenics. Traditionally, the term often stands for a coercive social program initiated by the state. Such a governmental eugenics might possibly also use modern gene technology. By using germline gene therapy the incidence of certain genetic diseases can be reduced in society, as we have seen in an argument in favor of germline gene therapy above. This new governmental eugenics might be coercive or voluntary. However, there may also emerge a kind of 'homemade eugenics' (Wright, 1990). In this case prospective parents choose the kinds of children they want. One could guess that most parents prefer healthy children and grandchildren, etc. This might lead some of them to request germline gene therapy.

It is not easy to assess the social risk arguments against germline gene therapy. This depends on the difficulty to anticipate future developments. The risks may be overestimated as well as underestimated. However, these arguments are the most important one against germline gene therapy. They each have a valid point, although they are not sufficient to reject all kinds of germline gene therapy.

The risks of discrimination, stigmatization and reinforcement of social injustice are real. Germline gene therapy should be accepted only if such bad consequences can be expected to be counteracted, at least to some extent. The key question is if this is possible. It should be noted that these risks are not unique to germline gene therapy. They are apparent also in somatic gene therapy, preimplantation selection, and genetic enhancement.

The slippery slope arguments should also be taken seriously, although the metaphor, as stressed in an earlier section, is not very good. A 'staircase' metaphor is preferable. But, of course, this metaphor opens up the possibility of stopping in time, and this would be contrary to the intentions of those using this argument. I will discuss enhancement below (including the argument about the possible use of germline gene therapy and enhancement by dictators) and eugenics in Chapter 6.

Other Kinds of Argument

There are also other objections to germline gene therapy. One focuses on the principle of informed consent. Above, I discussed this principle in relation to somatic cell gene therapy research. According to the present argument, germline gene therapy should not be accepted since the future generations affected by these interventions have not given their consent.

This argument is not acceptable. However, it is of some interest in relation to one of the main issues in the previous chapter, namely that of whether the concept of consent should be extended from individuals, to families and to communities. Since informed consent can be justified with reference to the principle of autonomy, the question about the extension of the concept of consent is closely linked to the question about the extension of the concept of autonomy. The argument at hand can be interpreted as an argument that the autonomy of the future people should be respected and that this implies obtaining their consent. If their consent to germline gene therapy cannot be obtained, germline gene therapy should not be carried out.

An obvious objection to this argument is that the principle of informed consent is irrelevant in this context, because future persons are nonexistent and therefore cannot give their consent. A more reasonable interpretation of the argument is that what its proponents in fact worry about is whether we can expect germline gene therapy to be in line with the interests of future generations. Would future persons, whomever they are, want us to use germline gene therapy or not? To be sure, it is difficult to predict the interests of future generations. However, we do know a lot about the interests of future persons. We can be very sure that they would prefer not to suffer from serious diseases. We have no reason whatsoever to assume otherwise.

Even if they might criticize germline gene therapy on moral grounds, they would not complain about not being born seriously ill.

Another objection to the argument is that it overstates the uniqueness of affecting future generations by germline gene therapy. Everything we do, not just germline gene therapy, affects the identity of the people who are born and in most cases we do not worry about this.

However, there is a difficult conceptual problem raised by this fact, which I have already discussed in general terms (see pp. 68-69). The key point is that since different decisions, reproductive or other, result in different future persons, there simply are no future persons who would be made better or worse off by either decision. For instance, whether we perform germline gene therapy or not no future persons would be better of worse off, because if we perform germline gene therapy other persons would exist than if we did not perform such therapy. One objection to this argument is that we are not morally responsible to any particular future persons but to the interests of those persons, whomever they turn out to be. Another objection starts from a distinction between 'possible people', those who could but do not necessarily exist, and 'future people', those who will exist in the future. It is absurd to say that we are morally responsible to merely possible people. But it is meaningful to talk about a moral responsibility to future people. The happiness and suffering of the latter can be compared. And it would be irresponsible to bring into being persons who will almost certainly be unhappy–they need a kind of minimum quality to their lives. Both these points support the argument above that germline gene therapy at least on serious diseases could be in the interest of future persons.

Another argument draws attention to the allocation of resources in biomedical research and health care. Germline gene therapy would be very expensive and would probably be so for a long time. Selective abortion and preimplantation selection are alternative options that are less expensive. Therefore, germline gene therapy will not for a foreseeable future be cost effective enough. A supportive and independent argument is that germline gene therapy is medically superfluous. In most cases, selective abortion or preimplantation selection are better alternatives. Affected zygotes can be discarded and nonaffected zygotes can be implanted. Germline gene therapy will also always imply more risks. Moreover, preimplantation selection has a clear advantage over selective abortion, since terminating the life of a fetus seems more dubious than discarding zygotes. On the other hand, embryo selection demands IVF, which for a foreseeable future will be very expensive. This makes selective abortion a more realistic option for many years ahead.

These arguments are very important. They make it probable that germline gene therapy as a means of treating certain diseases will be used only very rarely. In most cases, preimplantation selection would be a better option. But, to be sure, there are some circumstances under which germline gene therapy cannot be replaced by preimplantation selection. First, it might be requested when both parents are homozygous for the same recessive disease. In this case, all fetuses would be affected. Second, it might be an option when one or both parents are homozygous for a dominant disease. Even in this case the result would be that all fetuses are affected. In addition, germline gene therapy might be requested by parents who are both heterozygous for a recessive disease and are opposed to selective abortion and selective discard of zygotes for religious or other reasons. They might think that germline gene therapy is more in line with the general goal of health care of giving treatment than terminating life (Walters, Palmer, 1997, p. 85).

However, there is a possibility that germline gene therapy will be used as a means of treating diseases much more commonly than suggested above. Earlier I referred to the view expressed by certain geneticists and gene therapy researchers at the 1998 conference in Los Angeles that somatic gene therapy will never work particularly well but that germline gene therapy will. I do not share this pessimistic view of somatic gene therapy. But, on the other hand, if the optimistic view of germline gene therapy turns out to be true, the therapy might become a means of treating many rather common diseases.

Moreover, even if germline gene therapy would be used as a means of treatment only very rarely, it could be very useful in prevention. To eliminate a genetic disease for good in a family line would be a great medical benefit. But such genotypic prevention should only be accepted at the request of the prospective parents and it should not be a part of a public health program.

Conclusion

Germline gene therapy is a very controversial issue that the genetics community cannot ignore. It has to be discussed thoroughly in an open dialogue. As a contribution to this dialogue, I have carried out a rather systematic analysis of the main arguments. I conclude on the basis of this analysis that there are no completely decisive arguments against germline gene therapy but at least a few strong arguments in favor of it. This does not mean, however, that germline gene therapy research should be supported or even permitted. Germline gene therapy is acceptable only on certain very strict conditions. At the end of Chapter 6 I will clarify what these conditions are. To be sure, they are not at all fulfilled at present. For a foreseeable future, at least, germline gene therapy research should not be carried out.

3. GENETIC ENHANCEMENT

Let us now turn to genetic enhancement.[18] However, before we can start deliberating on the pros and cons of genetic enhancement, a cluster of conceptual problems has to be addressed (cf. Juengst, 1999).

Conceptual Problems

It is common to make a distinction between gene therapy and genetic enhancement. But how sharp is the line between therapy and enhancement? Therapy is often thought to have diseases as its targets, while enhancement has nondisease traits. This is, of course, a matter of stipulation but it seems quite reasonable to argue that enhancement can be disease-related. One possible example could be improvement of the immune system. As a result of genetic intervention people might acquire a highly above normal resistance against conditions like AIDS, or some other disease against which no humans have resistance. Another example could be to add cellular receptors to process cholesterol (Juengst, 1997). Genetic enhancement is often discussed without making the important distinction between disease-related enhancement and nondisease-related enhancement (Torres, 1997). This is not good, and not only for purely conceptual reasons. The distinction might also be ethically relevant.

Another conceptual problem, however, is whether only improvement of an ability from normal to above normal should count as enhancement or if improvement from below normal up to normal should as well. This is, of course, also a matter of stipulation. However, I would suggest that this latter improvement counts as therapy, not as enhancement.

I have already given a few examples of different kinds of disease-related enhancement. Nondisease-related enhancement, on the other hand, can also be of several different kinds. One example is cognitive improvement. Another is improvement of physical strength. A third kind is moral improvement of, for instance, sociopaths so that they get moral sentiments and become less aggressive. It also possible that parents might request enhancement of the aggressiveness of their children in order to help them in an increasingly competitive society.

Genetic enhancement can possibly be performed on somatic cells or on the germline. Both somatic enhancement and germline enhancement have been strongly criticized. However, the distinction between somatic and germline enhancement might be morally relevant. Perhaps somatic enhancement is acceptable, while germline enhancement is not?

Moral Imagination: Arguments

---

[18] For an early presentation of key arguments, see Glover, 1984.

All the arguments against somatic gene therapy can, with minor modifications, be used against somatic gene enhancement as well, and all the argument against germline gene therapy can with minor modifications also be used against germline genetic enhancement. Above I have found no decisive arguments against somatic gene therapy or germline gene therapy. They do not work as decisive arguments against disease-oriented genetic enhancement, either. But I shall not discuss these arguments here. Instead I shall focus on some other arguments directed more specifically against genetic enhancement.

One argument, already mentioned above in relation to slippery slope arguments, warns against possible misuse of genetic enhancement by dictators or totalitarian states. A popular version of the argument speculates about what could have happened if Hitler had had knowledge of genetic engineering techniques. By enhancing techniques he could have created elite soldiers whom no one could stop. Genetic enhancement is a dangerous instrument in the hands of a dictator. In my opinion, this argument is not particularly compelling. Genetic enhancement seems to be an unnecessarily complicated and expensive method for dictators and totalitarian states, at least for a foreseeable future.

Another kind of argument warns against a new eugenics. As mentioned above, the term traditionally refers to a social program initiated by the government. There is a possibility that genetic enhancement techniques could be used in such a program to genetically design the citizens. This new governmental eugenics might be mandatory or voluntary. But there might also emerge a kind of 'homemade eugenics' where prospective parents choose what kinds of children they want. This might lead some of them to request genetic enhancement, disease-related or nondisease-related. Children will be 'designed' to be, for instance, more resistant to disease, more intelligent, physically stronger or more 'physically attractive'. Moreover, such enhancement would probably for a foreseeable future be an opportunity only for the rich. We would get a 'new genetic upper class' born with far better potentials than other people.

This kind of an argument should be taken seriously. I shall discuss eugenics extensively in Chapter 6. I shall argue that nondisease-related genetic enhancement should be discouraged but that disease-oriented genetic enhancement should be allowed as a free choice of prospective parents, although not as a governmental program. This suggestion presupposes that it is possible to make a clear distinction between disease-oriented and nondisease-oriented genetic enhancement.

4. ANIMALS USED IN HUMAN GENETICS RESEARCH

The Human Genome Project includes not only an analysis of the human genome but also of the genomes of certain model organisms; for example, bacterium E. coli, yeast, round worm, fruit fly and laboratory mouse. The reason for mapping and sequencing such model organisms is to make possible a comparative analysis and increase the understanding of the human genome. The mapping and sequencing of the genomes of such model organisms is uncontroversial and poses in itself no serious ethical problems.

However, there are some controversial uses of animals in human genetics research as well that responsible geneticists need to consider. One example is the use of knock-out animals in the analysis of the functions of genes. Equally disputed is the creation of transgenic animal models for human diseases. Such models are used in order to understand the diseases and also to test different kinds of treatment, e.g. drug treatment or gene therapy. Moreover, transgenic animals are developed for drug production, and transgenic animal organs for human transplants are subjects of research. These uses are also rather controversial.

Many ethical questions arise in connection with such uses of animals in human genetics research. The general question is: what is it to take responsibility for laboratory animals used in human genetics research? Is it at all acceptable for responsible geneticists to use animals in this kind of research? If so, to what extent? Are there any ethical differences between using animals for basic genetic research and applied research? Are there any ethical differences between using transgenic animals and nontransgenic animals?

I shall discuss some of these questions. First, I shall analyze the ethical positions and arguments regarding the use of animals in research in general. This will constitute the main part of this section, and it is necessary in order to put the use of laboratory animals in human genetics research into a proper perspective. It also makes it possible to see whether this use differs from the ordinary use of animals in scientific experimentation. Second, I shall make a short comment on the ethical aspects of mapping and sequencing model organisms. Finally, I shall discuss specific problems related to the use of knock-out mice, the use of transgenic animals as human disease models, the creation of transgenic animals for drug production, and the use of transgenic animal organs for human transplants.[19]

The overall objective is to point out certain problems and arguments that responsible geneticists need to consider. The suggestions are only contributions to a dialogue, the genetics community has to formulate its own policy.

Ethics of Animal Experimentation: Two Perspectives

It is evident that the modern ethics of the use of animals in research has developed from two different sources. On the one hand, an ethical awareness of animal well-being has emerged from within science itself. A milestone in this development is the book *The Principles of Humane Experimental Technique* by William Russell and Rex Burch from 1959 (Russell, Burch, 1992 (1959). On the other hand, criticism of

---

[19] Note that when I talk about 'animals' I refer to nonhuman animals. To avoid this tiresome term, I use the shorter one as an abbreviation.

the use of animals in research has come from different Animal Rights movements with a base outside the scientific community. I will not try to determine here which influences are the most important. But it is obvious that a modern discussion requires a consideration of both perspectives. The use of animals in research is an activity carried out right in the middle of society. It is of great social concern. However, to the extent that we accept at least some use of animals in research, the more specific considerations about animal well-being in research must be made in intimate connection with practical scientific work. In this section, I shall analyze some of the main positions in the ethics of animal research. Some of the more specific issues will be discussed in later sections.

Major Positions

There are many different ways of categorizing the positions of the ethics of animal experimentation. I do not find it particularly enlightening or useful to start from abstract theorical constructions such as anthropocentrism, biocentrism, etc. These categories are much too many-faced and open to several different interpretations. Anthropocentrism, for instance, might include both the view that we may carry out any experiments we want and the view that we should be very restrictive. Similarly, biocentrism might include both the view that we should not carry out any animal experiments whatsoever and the view that we may carry out some animal experiments but not others. Thus, there might be an overlap between some versions of anthropocentrism and some versions of biocentrism. It seems more fruitful for my purposes to focus on positions regarding animal experimentation directly and on the arguments that are actually put forward. Most people accept some animal experiments but not others. Consequently, the important thing is to focus on particular kinds of experiment, not on abstract theory.

A problem in the dialogue on animal experimentation concerns the definition. In Swedish law, for example, the purpose determines whether an activity should be classified as an animal experiment (Swedish Animal Welfare Act, section 19). By contrast, the EU defines an activity as an animal experiment if it is an experiment for scientific purposes that involves harm to animals (Council Directive, 1986, Article 2). I shall not discuss the pros and cons of these two kinds of stipulation. Let me just state that I will exclusively discuss experiments that involve harm to animals. So when below I talk about animal experiments I refer only to experiments involving animal harm. But this should not be seen as a contribution to the debate on the definition of 'animal experiments'.

There are three major positions on animal experiments involving harm:

(1) We may carry out any animal experiments we want.

(2) We may carry out some animal experiments but not others depending on particular animal characteristics and particular circumstances.

(3) We may not carry out any animal experiments whatsoever.

Positions (1) and (3) represent extremes. Position (2) is intermediate. Rather few people adopt the first and third views. The first had its defenders, for instance, in Descartes, the famous 17th century philosopher, and Claude Bernard, the founding father of physiology. The third view is proclaimed by many animal rights activists, but not by all of them. The most radical are against all use of animals in research.

Most people take the intermediate course, i.e. position (2) (Orlans, 1993, pp. 20-34). This position encompasses, however, many different views, from those that are very restrictive, at the one extreme, to those that are very permissive, at the other. In essence, they all share the view that animal harm should be accepted only if vital human interests require it. But they differ in the opinion on what constitutes vital human interests. This means that the character of the dispute is quite in line with the general view on morality maintained in this book. There is agreement about the ethical principle of avoiding unnecessary harm to animals in research, but disagreement about its interpretation and application. There is a consensus about prototype cases of violation of the principle, but divergent views on how far the principle should be extended to nonprototypical cases. There is also disagreement about the justification of the principle. Some use deontological arguments, others consequentialist. The main problem within the second position is where to draw the line for acceptable animal experimentation.

Moral Imagination: Arguments in Favor of Animal Experimentation

In taking responsibility for the use of animals used in human genetics research, it is necessary that responsible geneticists use their moral imagination in considering a variety of arguments regarding animal experimentation in general. Let us start with the *pro* arguments. Why carry out animal experiments? There are at least three major reasons. The first refers to expected medical benefits for humans. This is, of course, the main reason. The second argument stresses medical benefits to animals. Some animal experimentation may lead to improvements for domesticized and other animals. The third argument points to the acquisition of basic biological knowledge. Commonly, animal experimentation aiming at increased basic knowledge is ultimately justified with reference to potential beneficial applications. But some would argue that gaining such knowledge would be reason enough.

However, it is not self-evident that the question should be posed in general terms. I will argue below that animal experimentation should be approached on a case-by-case basis. Scientists should ask in each particular case whether this or that specific animal experiment is justified.

Moral Imagination: Arguments Against Animal Experimentation

Let us now take a look at some arguments against animal experimentation, i.e. those in favor of position (3). If at least some of these arguments carry at least some conviction, I do not have to discuss the arguments in favor of position (1). As we shall see, I think that this is the case.

A fundamental problem with these arguments is that they can be used in favor of many different views. They can be used in favor of all the more or less radical views belonging to position (2), but they can also be used in favor of position (3). When they are used in favor of different views belonging to position (2), they are given weak interpretations. When they are used in favor of position (3), they are given strong interpretations.

There are many different ways of classifying the arguments. I find it useful to focus on the following four kinds of arguments.

(a) arguments from the integrity of animals,
(b) arguments from animal rights,
(c) arguments from sentience, pain or interests of animals, and
(d) arguments from the '*telos*' or species-normal functioning of animals.

Note that concepts like integrity, rights, or telos are sometimes related to one another in a particular argument. For example, animal rights may be thought to be based on the integrity of animals. The classification of the arguments focuses on what is put forward as the key ideas of the arguments. Sometimes, it may be possible for a particular argument to refer to more than one category, since many different concepts are used and there are different possible interpretations of what constitutes the key idea.

Below I will not give an exhaustive analysis of these arguments. This is not necessary for my purposes in this study. I will try to show that for each kind of argument there are strong versions as well as weak versions (cf. Moros, 1996). According to the strong versions, humans and animals have equal moral standing. According to the weak versions, they have unequal moral standing. The dignity, intrinsic value, rights, interests etc of humans always, commonly, or at least sometimes, outweigh those of animals. The strong versions impose strong restrictions on animal experimentation but are untenable, while the weak versions do not justify any strong restrictions but are more or less acceptable. This means that in their 'weak' versions these arguments can be viewed as arguments in favor of position (2), i.e. the intermediate position, which is the one I support.

Integrity

The first kind of argument centers around the concept of integrity.[20] Animals are wholes that should not be broken up and thus be violated. They have a dignity that should be respected. They do not only have an instrumental value but intrinsic value as well.[21] To paraphrase Kant, we should treat animals not as means only but as ends in themselves.

The arguments from animal integrity can be interpreted more or less radically. Defenders of the view that animals have dignity and intrinsic value can be positive as well as negative towards specific animal experiments. This depends on whether animals are thought to have equal or lower dignity/intrinsic value than humans. According to the 'strong' versions of the arguments, their dignity or intrinsic value is equal. In this case we would not be justified in carrying out experiments on animals that involve harm. A proponent of a strong version is Paul Taylor. He states that all individual living beings including animals have 'inherent worth' (Taylor, 1986, pp. 55, 63, 66-68, 75-77). On the other hand, if animals are considered to have lower dignity or intrinsic value, as the 'weak' versions of the arguments state, then we may carry out at least some of these experiments.

Animal Rights

The second kind of argument is stated in terms of animal rights. In the human context, rights can be understood in a strong sense and in a weak sense. Rights in a strong sense are correlated to obligations. Only creatures that have obligations–or belong to a community that has obligations although they themselves do not, for instance, newborn children–have rights. A right is a claim within a community of moral agents. It has a contractual character.[22] This is a prototype case of rights. However, many also accept noncontractual rights (Midgley, 1995). They are prepared to also extend the metaphor of rights to creatures that have no obligations. An example from the human context could be newborn children. This opens up the possibility to extend the metaphor of rights even to animals. This is what is done in arguments from animal rights.

Animal rights are justified by reference to certain animal characteristics. Let us take Tom Regan's view as an example. According to Regan, the relevant characteristic is whether a being is a 'subject of a life'. Such a being is psychologically complex enough to have some sense of itself as a being with a past

[20] For a recent analysis of the concept of animal integrity, see Heeger, 1997.

[21] In this discussion, many different terms are used: 'intrinsic value', 'inherent value', and 'inherent worth'. Some philosophers make clear differences in meaning between these terms, although they do this in different ways. In spite of the simplification, I will use 'intrinsic value' as a general term. This seems appropriate given my purposes in this chapter.

[22] A proponent of a contractual view of rights who criticizes the use of the concept of rights in relation to animals is Carl Cohen; see Cohen, 1994.

and future, some ability to act with a purpose, and some ability to suffer and benefit from the actions of others. Beings with such characteristics have 'inherent value' and this value is the basis for having rights. Regan argues that this is the case, not only of humans but also of many animals (Regan, 1988).

Even utilitarians may talk about animal rights, but then rights are not based on 'inherent value' but on a consequentialist analysis. To talk about animal rights can be useful in order to stimulate people to treat animals well.

The argument from animal rights can be interpreted more or less radically. As is the case with the arguments from integrity, it is also obvious that defenders of animals rights can reject as well as accept animal experiments. The key issue is whether animals are thought to have equal or lower rights than humans. The strong versions of the argument stress that their rights are equivalent (Regan). In this case, we would not be justified to carry out any experiments on animals that involve harm. According to the weak versions of the argument, animals have weaker rights. In this case, we may carry out such experiments to some extent.

Sentience, Pain, or Interests

The third kind of argument focuses on sentience, the ability of being conscious in the sense of being able to feel pain. Pain is here understood in the wide sense. It includes stress, distress, fear, anxiety, and discomfort. These arguments are commonly utilitarian but sentience and pain might, of course, be considered ethically relevant according to other views as well. The utilitarian arguments can be articulated in terms of pain and pleasure (classical utilitarianism) or in terms of interests (interest utilitarianism). The most well-known present-day proponent of an interest utilitarian argument against animal experimentation is Peter Singer. His key idea is that if those who want to carry out animal experiments are not prepared to use a human infant or a brain-damaged person, they reveal an unjustifiable form of discrimination on the basis of species. Such a 'speciesism' would be analogous to racism. However, Singer is not in favor of an absolute prohibition. There might be some extremely rare cases where it would be possible to save thousands of human lives by an experiment that would take just one animal life. They would be justified. But this is not at all the case in ordinary animal experimentation (Singer, 1990, pp. 80-87).[23]

As with the preceding arguments, the arguments from sentience, pain or interests of animals can be interpreted more or less radically. There are strong versions according to which the pleasure, pain, and interests of animals rank equally with human pleasure, pain or interests. Singer comes close to such a view. However, he talks about 'equal consideration' of interests, not of equal moral standing. Moreover, there are weak versions according to which animal pleasure, pain or interests are

---

[23] See also Rachels, 1990.

ethically relevant but lower ranking than human pleasure, pain or interests. According to the weak versions, we may carry out some animal experiments but not others. A proponent of a rather weak, but not very weak, view is Bernard Rollin (Rollin, 1995, p. 172).

Telos and Species-Normal Functioning

Finally, there are arguments pointing at *telos* or purposiveness. It is important here to make a distinction between *telos* in an Aristotelian sense and *telos* in a Darwinian sense. *Telos* in the Aristotelian sense is essentialist, referring to the pregiven essence or nature of an organism. *Telos* in the Darwinian sense is antiessentialist, characterized by continuity-through-change. This means that '*telos*' refers to the normal or natural life at a given point in the evolution of a species as a result of natural selection. Modern arguments from *telos* are commonly based on an evolutionary understanding. Their point is that we should not do things to animals that hinder their possibility of leading a life in accordance with their species-normal functioning. In this way, other aspects than the subjective experience of feeling well can be considered ethically relevant, namely functioning well and leading a natural life. And this may hold for sentient as well as nonsentient animals.

Even these arguments can be interpreted more or less radically. The strong versions stress equal consideration of the species-normal functioning of animals and humans. A defender of this kind of view is Paul Taylor, who I have already presented above as a proponent of the view that all living beings including animals have an 'inherent worth'. The weak versions, on the other hand, maintain that while the species-normal functioning of animals is ethically relevant, the species-normal functioning of humans ranks higher. This means that some animal experiments may be carried out, while others should not. A proponent of such a view is Arthur Caplan (Caplan, 1993).

Discussion: Equal or Unequal Moral Standing?

In all these different kinds of argument we have found a common pattern. Each argument can be interpreted more or less radically. There are strong interpretations as well as weak ones. This is precisely what could be expected from the point of view of imaginative casuistry. In the following discussion, I shall restrict myself to some general comments on these interpretations, and ignore the more specific problems that might be connected with each particular argument.

Let us start with the strong versions. In these versions, the rights, interests, etc. of animals and humans have an equivalent standing. Animals and humans have equal moral standing. The fundamental problem with these versions is how to adjudicate competing rights, interests, etc. What should we do when the rights, interests, etc. conflict? Take the example of a dog attacking a little child. Should you, if necessary in order to save the child's life, shoot the dog? If so, you have to be prepared to do the same if the dog attacks a squirrel, otherwise you make an ethically relevant distinction between a human and an animal, and consequently you are not holding to the strong interpretation. Furthermore, a consistent application of this kind of idea seems to imply that we should be at constant war with predators. But this would lead to ecological disaster. Moreover, this would conflict with our duties to the predators. How can we be at war with predators and at the same time treat them well? In conclusion, it appears that the strong versions of the arguments are untenable.

What about the weak versions? These versions have the common point that the rights, interests, etc. of animals are weaker than those of humans. Animals and humans do not have equal moral standing. Animal rights, interests etc should be respected but vital human rights, interests, etc. may outweigh animal rights, interests, etc. This implies that there is no strict prohibition of animal experimentation. It is possible that human rights or interests outweigh animal rights or interests. It becomes possible to perform certain animal experiments for the benefit of humans even if this implies a violation of animal rights or causes animal pain.

It is obvious that the weak versions of the arguments do not achieve what some animal activists would wish, namely to prohibit or at least very strongly restrict animal experimentation. They are not tenable as arguments in favor of position (3). On the other hand, the weak versions do speak against position (1). There are at least some ethical restrictions on animal experimentation. Humans are not allowed to carry out any animal experiments they want. To the extent that animals can be harmed, they should be subject to at least some moral consideration. This means that the weak versions of the arguments against animal experimentation can be used in favor of position (2), the intermediate position. However, they do not give specific guidance about which human interests are vital enough to justify animal experimentation. Do the rights, interests, etc. of humans commonly or only sometimes outweigh those of animals? The weak versions do not provide clear guidance. This means that they are of limited practical value.

A Shift of the Burden of Proof

As mentioned above, I will not give a specific assessment of each of the arguments. However, it is necessary to make a comment on the possibility of harming animals. At least to the extent that animals are sentient and can feel pain, they can be harmed. This fact must be the starting point of all reasoning about the moral acceptability of animal experimentation. It is uncontroversial. What is controversial is what we should make of this fact. The strong versions support the view that we should never do any animal experimentation whatsoever. The weak versions allow that at least some animal experiments might be acceptable. But they are of only limited help in drawing a strict line for acceptable use.

What is truly important, however, is that the weak versions of the arguments shift the burden of proof to those who want to carry out animal experimentation. They have to be able to show that the experiments are necessary even though they do imply some harm to animals.

Some Restrictions

The weak versions of the arguments put at least some important restrictions on animal experimentation. In general, they require that we should not cause animals unnecessary harm. They justify this requirement in divergent ways. Some of these justifications might not be tenable but I will not discuss this here. Defenders of all these different arguments agree that we should not cause animals unnecessary harm. More specifically, the weak versions require:
- that scientists and keepers should treat animals well in their ordinary life before and after specific experiments,
- that animals should be anesthetized if the experiments include interventions that might cause pain,
- that killing should be painless, if killing is necessary,
- that scientists, when possible, should reduce the number of animals used,
- that scientists, when possible, should refine the experiments in the sense that we should try to get as much scientific information as possible from each experiment and in this way reduce the pain of the animals, and
- that scientists, when possible, should replace animal experiments with other kinds of experiments such as experiments on cell cultures.

Note that the last three requirements are the 'three Rs' in animal research: reduction, refinement, replacement (Russell, Burch, 1992 (1959)).

A Shift of Focus to Practical Issues

A general conclusion of this analysis is that the weak versions make us aware of the general moral standing of animals. Scientists should not cause animals unnecessary harm. The arguments put some restrictions on animal experimentation, and shifts the burden of proof to those who want to carry out animal experiments. However, they are of limited help in practical work. As a result of this analysis, it becomes obvious that what is needed is a shift in the way the questions about the treatment of animals are put. Most people accept that we should not cause animals unnecessary harm and at the same time they accept that some animal experiments are necessary due to vital human interests. Two questions come to the forefront:
(1) When are human interests vital enough to justify animal experimentation?
(2) Given the fact that animal experiments will be a part of society for a foreseeable future, how can we reduce animal harm in animal experimentation?

These questions mark a shift in the way the questions about animal experimentation are put. Practical issues are of the most central importance. We have to decide in each particular case whether the human interests are vital enough and we have to decide in the particular context of each animal experiment what could be done to avoid or reduce animal harm. Only by such a casuistic and context-sensitive approach can scientists assume moral responsibility in animal experimentation.

It should also be noted that even if we answer "Never!" to the first question, moral responsibility demands that we make the second question our concern. Even radical animal rights activists should accept the fact that animal experiments will be a part of society for a foreseeable future. They cannot stop this. So, for all parties–activists, researchers, the general public–the second question becomes truly important.

Genetic Engineering and Animal Well-Being

I have analyzed and discussed some key aspects of the ethics of animal experimentation in general. I have defended the view that it is acceptable to carry out at least some animal experiments. Now, I shall turn to the ethics of genetic engineering of animals. More precisely, I shall discuss the use of transgenic animals in human genetics research (cf. van Zutphen, van der Meer, 1997).

Let me first point out a conceptual problem. The term 'transgenic animal' refers to an animal whose genome has been deliberately modified by the transfer of foreign DNA to every cell, including germ cells. Commonly, the term also includes a 'knock-out' animal, i.e. an animal in which a gene has been inactivated (Mepham et al., 1998). The terminology, however, is not quite strict. I will use the broad definition.

Examples of the use of transgenic animals in human genetics research are: knock-out animals in the analysis of the functions of genes, transgenic animal models of human diseases, transgenic animals for drug production, and transgenic animal organs for human transplants. A central ethical issue for responsible geneticists to consider is, of course, if it is ethically acceptable to genetically modify animals for use in human genetics research. But it also vital to consider more in detail whether there are any ethical differences between different kinds of uses.

Let us start with a presentation of some arguments for and against genetic modification of animals for use in human genetics research that responsible geneticists need to consider.

Moral Imagination: Arguments For and Against Using Transgenic Animals in Human Genetics Research

The main *pro* arguments stress the medical benefits and the value of new basic biological knowledge gained through this research. This is actually the same arguments as the general ones in favor of animal experimentation.

Similarly, the general arguments against animal experimentation can also be used against genetic modification of animals for use in human genetics research. However, I shall not present these arguments once again. Let me just stress the importance of arguments focusing on the pain and suffering of animals used in research. It is obvious that genetically engineered animals might be subject to such pain and suffering.

Instead I shall focus briefly on two special arguments. They are quite similar to a couple of arguments against human genetic modification discussed earlier in this chapter. The arguments are based on the principle of integrity. According to the first, genetic engineering of animals for use in human genetics research would be to 'play God'. We are doing what only God should do. According to the second argument, such engineering is unnatural, a violation of the principle of respect for the natural order. Sometimes the metaphor of Frankenstein's monster is used in this kind of argument.[24]

Discussion and Proposal

I have referred to the European survey of public attitudes to different kinds of genetic application (see p. 87). An important finding was that the main reason behind the resistance of large parts of the general public had to do with the violation of the natural order and the integrity of Nature rather than with risks. Genetic modification of animals was viewed as morally unacceptable in itself, not only as dangerous (Biotechnology and the European Public Concerted Action Group, 1997). Geneticists should not underestimate the importance of this kind of argument among the general public. In order to gain credibility they have to take it seriously and try to put forward a strong argument against it.

In the discussion of human gene therapy, I argued that this kind of an argument has many weaknesses. The main objection is that the principles to which they appeal are ambiguous. These principles can be extended in many different ways. They can be used more or less radically. Moreover, some are based on certain life views that are not generally shared. If these life views are not accepted, the arguments carry little weight. As in the case of human gene therapy, this makes them no good basis for the professional policy of the genetics community.

However, it is not enough for scientists to refute the argument from the natural order. They also have to show that vital human interests are promoted by the use of genetically modified animals in human genetics research. A key question is: when are human interests vital enough?

---

[24] Cf. the title of Rollin, 1995: *The Frankenstein Syndrome*.

Of the examples mentioned above, the one most easy to accept is probably the creation of transgenic animals for drug production. A well-known example is the sheep Tracy, which has been genetically engineered to produce human insulin. Other examples are transgenic animals used for the production of human growth hormone (Reiss, Straughan, 1996, p. 166). These animals do not suffer due to genetic engineering. They live as good lives as other domestic animals. In fact, they might even lead better lives since they are considered to be of great economic value and consequently are treated very carefully. Therefore, there is, in my opinion, no reason to reject this application.

Somewhat more problematic is the use of genetically modified animal organs for human transplants. The pig Astrid is a well-known example (Reiss, Straughan, 1996, p. 173). The main argument for xenotransplantation, i.e. using animals for human transplants, is that other alternatives do not seem not sufficient to counteract the shortage of organs or in some cases even acceptable. One alternative is the willingness of people and relatives to altruistically donate organs.[25] This is the traditional method. Another option is to allow a free market for organs (Blumstein, 1996). A third possibility is to enact laws about presumed consent to donate organs after death. People are not, as is commonly thought, presumed not to consent but to consent. This was stated in the Swedish Transplantation Act from 1996. Finally, there is the alternative of constructing artificial organs. The first alternative does not seem sufficient. The second is very controversial. The third has not as yet led to an increased number of available organs, and the fourth has had at least some success.

With this in mind, xenotransplantation might be an important solution to the serious problem of shortage of organs. A medical obstacle is the immunological rejection of alien organs in the human body. This problem is thought to be solved by genetic engineering. One alternative could be to transfer human genes to the pig resulting in a transgenic pig. Another alternative could be to 'knock-out' animal genes, i.e. remove the genes or make them inactive, in order to prevent rejection.

Is xenotransplantation using transgenic animal organs for human transplants ethically acceptable? Some find it unnatural. On the other hand, to patients with no other option but death, it might appear necessary. In my opinion, this is decisive. The principle of beneficence requires that we find treatment for these patients. A scientifically important problem, however, is that the use of genetically modified animal organs might spread pathogens. We should take this threat seriously but not overestimate it.

Moreover, transgenic animals may also be used in order to understand human diseases and develop treatments. Particularly in immunological research and the study of cancer development transgenic disease models may be of great importance. A well-known example is the 'Harvard oncomouse'. Other examples are animal models for diseases like AIDS, SCID, and Lesch-Nyhan Syndrome (Reiss,

---

[25] For a general discussion of altruism in society and health care including organ donation, see Nordgren, Westrin (eds.), 1998.

Straughan, 1996, pp. 169-171). Human genes for these diseases are transferred to animals, and lines of transgenic animals with the diseases are created. These animal models are then used for the development and testing of drugs and other therapies. It is also possible to make knock-outs for use as disease models.

The use of transgenic animals as human disease models is more problematic than the uses of transgenic animals as drug producers and 'donors' of organs for human transplants. One objection is that the results to date have been poor. The animals suffer and the medical benefits to date have been very limited (Reiss, Straughan, 1996, p. 191). On the other hand, if there is a breakthrough the assessment might change. The animal experiments will be seen as worthwhile. We have to make a case-by-case judgment concerning this application. The scientific anticipation of future success in each particular case will be of key significance.

The examples of the use of transgenic animals discussed so far have all been related to the understanding and treatment of human disease. What about the use of animals for obtaining pure basic genetic knowledge? This seems more problematic, since we do not know in advance whether this basic knowledge will be medically useful. On the other hand, we know from the history of science that many basic findings with no anticipated application have actually led to applications of great medical and other benefit.

In human genome research, the use of knock-out mice and rats is widespread. They are used in order to obtain basic knowledge of the functions of genes including disease-related genes. A particular gene is knocked-out in order to find out its specific function. Such experiments may lead to pain and suffering of the animals. Is this ethically acceptable? Even in this kind of research, we have to make a case-by-case assessment as suggested by imaginative casuistry.

Conclusion

In conclusion, the uses of transgenic animals for drug production and as suppliers of organs for human transplants are not particularly problematic. The use of transgenic animals as disease models is more problematic. This holds true for the use of knock-out animals in function analysis as well. Both uses can, however, be accepted in principle. But the experiments must be judged on a case-by-case basis.

Responsible geneticists need to make their own ethical deliberation. However, if they accept the above considerations, it seems vital that they make particular efforts to design these two kinds of experiments carefully. For reasons of animal welfare, they need to reflect thoroughly on the justification of their experiments.

CHAPTER SIX

EUGENICS AND THE LONG TERM GOALS OF REPROGENETIC MEDICINE

1. RESPONSIBILITY FOR APPLICATIONS OF GENETICS: PROFESSIONAL POLICY

The *Declaration of Inuyama* stresses very clearly the social responsibility of scientists doing human genetics research. Two of its key statements are as follows (CIOMS, 1995):

> In terms of ethics and human values, what must be assured are that the manner in which gene mapping efforts are implemented adheres to ethical standards of research and that the knowledge gained will be used appropriately, particularly in genetic screening and gene therapy...
> Genetic researchers and gene therapists have a strong responsibility to ensure that the techniques they develop are used ethically. By insisting on truly voluntary programmes designed to benefit directly those involved, they can ensure that no precedents are set for eugenic programmes or other misuse of the techniques by the State or by private parties.

I shall take these statements as a starting point for a discussion of the responsibility of geneticists for the societal application of their findings. The knowledge obtained through the human genome research might be used, for instance, in genetic testing, genetic screening and gene therapy, and the *Declaration* stresses strongly the responsibility of the geneticists for such applications.

However, do geneticists conducting genetics research truly have a responsibility for the uses other scientists, industrialists or politicians make of their results? This is not self-evident and some would say that geneticists do not have such responsibility. Let us therefore repeat the arguments in favor of this view. I have put forward two general arguments for the view that scientists should take responsibility for trying to affect the applications of their findings made by other people. Applied to geneticists, I have argued:

(1) that geneticists have a duty to respond to the needs and fears of the general public, because their research has a social impact, and

(2) that geneticists owe this concern to society, because they have received from society education, resources, and access to the general infrastructure.

To these arguments I have added two further ones that are particularly applicable to geneticists:
(3) that geneticists can often anticipate what possible applications might be made, and
(4) that at least to the extent that certain possible applications are referred to as a justification when applying for grants to basic research, geneticists should take responsibility for these applications.

Given this view that geneticists should take responsibility for affecting applications of their findings made by other people, what are the implications? What should the geneticists do? From the perspective of individual geneticists and research teams, it is vital to reflect on the research projects at hand and their possible applications. The geneticists should reasonably ask themselves: Are we proud of what we are doing? Are there any possible applications of this particular research which would not be ethically acceptable? What should we do in order to prevent that these applications are realized? What should we do in order to promote good applications? The individual geneticists and research teams

From the perspective of the genetics community at large, it is essential to try to develop a common professional policy. This policy should define genetics as a science and its role in society. In order to do this, the geneticists have to use their scientific and moral imagination to envision the long term goals of genetics. Alongside the purely scientific goals, it is vital that they consider the use of scientific findings in medicine and society at large. What should be the goals of this medical and societal use? Which applications are acceptable given these long term goals? This is the issue of the content of responsibility. Both with regard to research and application it is also important to take a stand with regard to the means of implementing the goals. This is the issue of the forms of responsibility. To be sure, the formulation of a professional policy is in itself a possible form of responsibility. It is a way of taking responsibility in genetics. But such a policy should also include statements about more specific forms of responsibility. How should the goals of the policy be realized? Would it be necessary with legal regulation or would professional guidelines be sufficient?

Different branches of human genetics may have different tasks in articulating the professional policy. The goals of genetic medicine are, of course, primarily the task for medical geneticists and other geneticists working in the field of medicine. If geneticists cannot reach a general consensus on every issue–which, of course, is very unlikely–they should at least discuss the issues and try to reach whatever consensus is possible. A dialogue within the genetics community on these matters would be of great value, even if full consensus is not achieved. A dialogue is needed nationally as well as globally.

Why is it essential with a professional policy on these matters? Why cannot the genetics community just accept governmental policy on the use of genetics in medicine and society at large? There are at least two important reasons. First, it is a question of professional autonomy and self-understanding. The genetics community must envision and decide for itself what should be the goals of their own science and its role in society. It cannot just leave it to the government to determine this. A second reason is that scientific research often is one step ahead of politicians and legislators. Scientific achievements often occur without being anticipated by politicians and legislators. This holds true also of applications of the scientific findings. Scientists, on the other hand, are often in a better position to anticipate future developments in their own field. To the extent this is the case, scientists may also get a start over other people in anticipating possible applications. This means that a well-considered professional policy of the genetics community is necessary while waiting for a governmental policy. It may also give geneticists a unique opportunity to influence governmental policy. Precisely by being a step ahead of the government, the genetics community may provide valuable input to public policy and possible legal regulation.

Below I shall discuss the possible goals of genetics. But I shall have a rather restricted objective. I shall not discuss the goals of the societal use of genetics in general, neither the goals of genetic medicine in general. I shall focus on the long term goals of reprogenetic medicine, i.e. reproductive applications of genetics made in the medical context. One reason for focusing on reproductive applications and leaving out nonreproductive applications is that reproductive applications have more far-reaching implications for human life than the nonreproductive ones. They concern what many people consider to be the key issue in life, namely procreation and parenthood.

Another reason for focusing on reproductive applications and leaving out nonreproductive applications is historical and linked to the issue of eugenics that precisely concerns reproduction. Many geneticists in the early decades of the 20th century thought that social responsibility in genetics implied eugenics. Today social responsibility is usually understood as implying an obligation to avoid eugenics, as can be seen in the *Declaration of Inuyama*. The *Declaration* stresses the importance of "... truly voluntary programmes designed to benefit directly those involved", and makes a distinction between such programs and 'eugenic programmes' (CIOMS, 1995). It seems to reserve the term 'eugenics' only for nonvoluntary, coercive social programs. However, some proponents of eugenics at the beginning of the 20th century as well as modern proponents such as the philosopher Philip Kitcher emphasize the voluntary character of their eugenic social programs (Kitcher, 1996). It seems quite possible to talk about 'voluntary eugenics'. Moreover, if we use an even wider definition of 'eugenics', we might also talk about 'homemade eugenics' (Wright, 1990) as a possible consequence of the "truly voluntary programmes" in the *Declaration's* sense. Homemade eugenics is a kind of eugenics carried out by prospective parents. All this indicates that the concept of eugenics has to be analyzed thoroughly. It is also necessary to discuss different versions of eugenics and try to establish whether there are any morally relevant distinctions between these different versions. Should eugenics in one sense or another be a prototype for the professional policy of the genetics community?

2. DIALOGUE ON THE LONG TERM GOALS OF REPROGENETIC MEDICINE

In order to develop a professional policy on the long term goals of reprogenetic medicine, a thorough dialogue within the genetics community is necessary. Moreover, in considering these long term goals, it is particularly essential that responsible geneticists use their scientific and moral imagination and take into account a variety of possible perspectives and arguments. A consideration of the long term goals requires an ability to anticipate future scientific developments and their possible consequences. It also requires a determinate will to influence the development of reprogenetic medicine in a particular direction.

In order to place the discussion of reprogenetic medicine and eugenics in a proper perspective, it is necessary to briefly focus on the goals of medicine in general. In doing so it is vital to keep in mind the distinction between short term and long term goals. For various reasons, the short term goals of reprogenetic medicine may perhaps differ from the long term ones. Below I shall primarily focus on the long term goals.

The Goals of Medicine

It might be surprising that the issue of the goals of medicine in general is a largely forgotten issue. It has, however, far-reaching practical implications. In order to compensate for this shortage, an international project was initiated in 1993 by the Hastings Center and led by Daniel Callahan. A book, *The Goals of Medicine*, has recently been published, which includes a general Project Report as well as a wide range of contributions by some of the participants. The key idea is that the dominant 'diagnose-and-treat' paradigm is unsatisfying. The Report maintains that the following goals are important:
(1) disease prevention and health promotion,
(2) relief of pain and suffering caused by maladies,
(3) care and cure, and
(4) avoidance of premature death and pursuit of a peaceful death (Hanson, Callahan, 1999, pp. xi, xiii, 18-30).
I agree. However, it is interesting to note that the Project Report does not include individual autonomy and fair access to medical facilities as goals of medicine. It seems that it understands these values only as principles for the implementation of the goals of medicine or perhaps as secondary goals, not as primary ones. However, this is not uncontroversial. It constitutes a problem that has to be discussed.

Another problem with the Report is that it does not take seriously enough the potential conflict between a public health approach and a client-centered approach. The Report states that (Hanson, Callahan, 1999, p. 46):

> Public health programs should bear a large part of the task of health promotion but they can hardly do so without the active cooperation of physicians interacting with individual patients; both approaches, population- and individual-oriented, are necessary and can helpfully reinforce each other.

This optimism is somewhat moderated further ahead in the Project Report where the possible tension between indvidual good and social good is acknowledged. Individual freedom of choice is recognized but at the same time it is stressed that certain moral responsibilities goes with freedom (Hanson, Callahan, 1999, p. 52):

> Modern medicine presents a complex range of choices, many exceedingly difficult, to individuals and societies. A necessary moral condition for responding to these choices is democratic participation in societal decisionmaking, and the freedom of choice wherever possible in the individual decisionmaking. Freedom of choice, the fundamental right of self-determination, carries with it attendant duties and responsibilities... This is only to recognize the necessary and fruitful interaction–and sometimes tension–between individual good and social good.

However, the tension between population-oriented and individual-oriented approaches should be taken more seriously. What happens if individuals do not take their moral responsibility in the Report's sense? Should their free individual choices not be respected in that case? As we shall see below, the potential conflict is particularly apparent in reprogenetic medicine.

Moral Imagination: Possible Goals of Reprogenetic Medicine

According to the Project Report, disease prevention is one of the neglected goals of medicine. This possible goal is particularly important to discuss in relation to reprogenetic medicine. The other neglected goals, however, are less central, although they are very important in medicine in general. Another possible goal not mentioned in the Report but relevant to reprogenetic medicine is genetic enhancement. Two further possible goals are, as indicated above, individual autonomy and fair access. In sum, together with the traditional goals of diagnosis and treatment there are at least six possible goals for responsible geneticists to consider:

- diagnosis and prognosis,
- treatment and, if possible, cure,
- prevention,
- enhancement,
- reproductive autonomy, and
- fair access.

Diagnosis and Prognosis?

Diagnosis and prognosis in reprogenetic medicine might seem to be an unproblematic goal. However, as we have seen in Chapter 4, this is not the case. Let me briefly point out three important problems.

One problem concerns what kind of genetic information should be sought: information about very severe diseases, severe diseases, less severe diseases, minor afflictions, disease susceptibilities, late-onset disorders, or even information about nondisease traits such as sex (in the absence of sex-linked genetic risks) or intelligence (if possible in the future). It might seem that genetic information about nondisease traits fall outside the domain of reprogenetic medicine but this is not self-evident. Perhaps the domain should be expanded at the request of the clients?

Another problem concerns the concept of 'prognosis'. By talking about 'predictive' and 'presymptomatic' genetic testing there is a risk that we underestimate contingency. It is not always a realistic goal to make an accurate prognosis.

A third problem is raised by the therapeutic gap. This gap between the amount of genetic information we can obtain and the not too uncommon absence of proper prevention and treatment makes the goal of diagnosis and prognosis less obvious than it might first appear.

Treatment and Cure?

Treatment and cure by therapy might also seem an ethically unproblematic goal of reprogenetic medicine. But this is not true, either. The use of genetically tailor-made drugs is rather uncontroversial, as is somatic gene therapy. But, as we have seen, germline gene therapy is very much disputed.

However, there is not only a dispute regarding specific technologies. There are also different opinions about which diseases or traits these technologies should be applied to. Moreover, there is a problem concerning who should be considered to be patients. Is it only individuals, or also families and populations? The possibility of germline gene therapy also raises the question whether perhaps even future persons could be considered potential patients. By using germline gene therapy a specific genetic disease might be eliminated from a particular family line. This, however, implies that we slide over from treatment to prevention.

An even more fundamental problem concerns the concept of therapy itself. Karen Lebacqz argues that this concept in itself is ambiguous. What is 'therapy' to the dominant group in society may be an undesired intervention to marginalized groups such as blind people, deaf people, or people of very short stature. According to Lebacqz, disability is a social construct. What is considered a 'disability' from the dominant culture is not necessarily a disability to those who deviate from the cultural norm. Moreover, Lebacqz stresses that even where intervention is accepted it might be ambiguous. The 'self' that emerges following therapy may not be altogether comfortable. The sense of personal identity may have a tendency to remain with the untreated self. Lebacqz suggests that instead of 'gene therapy' we should talk about 'genetic intervention' (Lebacqz, 1999).

Others would say, however, that reprogenetic medicine would be too neutral if it gives up its therapeutic goal. The goal should be treatment and cure by therapy. But therapy should be provided only to those who want it. This means that the ambiguity of the concept of therapy can be used as an argument for a client-centered framework in contrast to a public health approach. In order to implement public health goals, reprogenetic medicine should not be overly focused on actively looking up potential patients.

Prevention?

Disease prevention is another important goal, according to the Project Report from the Hastings Center. However, in reprogenetic medicine the concept of prevention is problematic. The distinction between phenotypic and genotypic prevention is essential (Juengst, 1995b). This distinction is not made in the Report. Actually, these two forms of prevention represent two different possible goals for reprogenetic medicine. Phenotypic prevention is an uncontroversial goal in public health work. Genotypic prevention, on the other hand, is highly controversial.

Genotypic prevention can be a goal for reprogenetic medicine as well as for prospective parents. It is, therefore, important to distinguish between the following two meanings:

(1) genotypic prevention as a public health measure aimed at the reduction of the incidence of certain genetic diseases in the population, and

(2) genotypic prevention as a measure of prospective parents aimed at avoiding giving birth to children with certain clinical genotypes (cf. Juengst, 1995b).

Genotypic prevention in the first sense is a possible long term goal for reprogenetic medicine. In this sense it could be a key concept in a kind of 'voluntary public health eugenics'. It is a policy that is explicitly aimed at improving the health of future generations. Genotypic prevention in the second sense is a possible goal for prospective parents. It is a key concept in a kind of 'parental eugenics'. It should be pointed out that it is possible to accept genotypic prevention as a reproductive choice of prospective parents, while at the same time reject it as a public health measure.

Enhancement?

A fourth possible goal is genetic enhancement. This kind of genetic intervention is very much disputed. Could genetic enhancement in any sense and to any extent be an acceptable goal of reprogenetic medicine?

To some, this depends on whether it is possible to make a reasonably clear distinction between disease-oriented and nondisease-oriented genetic enhancement. If this is possible, disease-oriented genetic enhancement could be an important goal. Possible examples are, as mentioned in the previous chapter, improvement of the immune system, improvement of the efficiency of DNA repair, and adding cellular receptors to capture and process cholesterol (Juengst, 1997).

Reproductive Autonomy?

Respect for reproductive autonomy of prospective parents is a well-established principle in clinical genetics that is expressed in the policy of nondirective genetic counseling. This policy is traditionally thought to be an antidote to eugenics. An important question is whether reproductive autonomy should be understood only as a principle governing the implementation of the goals of reprogenetic medicine or as a goal in itself. Another important question is whether full reproductive autonomy, increased but not full reproductive autonomy, or just a possibility to accept or refuse recommendations from the government or the professionals of reprogenetic medicine is desirable.

Moreover, reproductive autonomy as a goal or guiding principle of reprogenetic medicine has two possible aspects, a negative and a positive. The negative aspect is freedom from coercion or pressure from other agents. The positive aspect is the provision by the government of certain means to realize the wishes of prospective parents. Should both aspects or only the negative one be accepted?

Fair Access?

Another important ethical principle is justice. When applied to reprogenetic medicine it may imply fair access to genetic facilities. Government should see to it that the benefits of reprogenetic medicine not merely become a privilege for the rich. An important question is whether fair access should be understood only as a principle governing the implementation of the goals of reprogenetic medicine or as a goal in itself. In Swedish health care legislation, for instance, care on equal terms for all is explicitly proclaimed as a goal. On the other hand, the Project Report while talking about a "just and equitable medicine", it seems to understand fair access as a principle governing the implementation of the goals of medicine, not as a goal in itself. An alternative interpretation is that it is a secondary goal, not a primary one.

Different Interpretations and Rankings of the Goals

All the above mentioned goals seem at least to some extent acceptable. But our analysis has indicated that they can be interpreted and balanced in different ways.

Actually, this is what could be expected from the point of view of imaginative casuistry.

According to imaginative casuistry, it is also important to recognize the framing of moral problem situations. In discussing the goals of reprogenetic medicine, it is vital to decide within what kind of framework the assessment of the realization the goals should be carried out. When they conflict, which framework should be prioritized, a public health framework or a client-centered framework? In relation to what framework should we measure success in reprogenetic medicine? As stressed above, the Project Report underestimates the possible tension between these two frameworks. This tension is particularly obvious in reprogenetic medicine.

Moreover, the tension between a public health framework and a client-centered framework is clearly related to the two possible goals of prevention and reproductive autonomy. Therefore, I shall discuss these two goals more in detail. With regard to prevention, I shall focus in particular on genotypic prevention, a very controversial issue due to the history of eugenics. With regard to reproductive autonomy, I shall pinpoint different interpretations, some of them radical, others more restrictive. Most geneticists would accept reproductive autonomy as an important value, but how far-reaching a reproductive autonomy should prospective parents be allowed?

3. LEARNING FROM HISTORY: EUGENICS

The question of the long term goals of reprogenetic medicine is closely related to the question of eugenics. Should the goals be eugenic in some sense? If not, what are the reasonable alternatives? The answers to these questions depend on what is exactly meant by 'eugenics'. As quoted at the beginning of this chapter, the *Declaration of Inuyama* stresses the importance of insisting on "truly voluntary programmes designed to benefit directly those involved" and on ensuring "that no precedents are set for eugenic programmes or other misuse of the techniques by the State or by private parties" (CIOMS, 1995). The *Declaration* clearly links the problem of the long term goals to the issue of eugenics and explicitly rejects eugenic programs. In the opinion of the *Declaration*, the long term goals should not be eugenic. However, eugenics can be understood in many different ways and perhaps there are ethically relevant distinctions between these versions. In any case, it becomes essential to analyze the term 'eugenics' more precisely. In conclusion, we have to deal with the question of eugenics if we want to have a broad discussion of the long term goals of reprogenetic medicine.

The Diversity of Eugenics

The term 'eugenics' was coined in the year 1883 by Francis Galton, the cousin of Charles Darwin. To Galton eugenics was "the study of the agencies under social control that may improve or impair the racial qualities of future generations, either physically or mentally."[1] Note that Galton here focuses on scientific study, not on social practice. The latter, however, is what the term usually refers to. Furthermore, it often stands for a coercive social practice initiated by the government. This can be seen, for instance, in the Declaration of Inuyama. In this sense it is heavily emotionally loaded. Everyone associates the term with Nazi racial laws and the Holocaust. However, some eugenicists at the beginning and the middle of the 20th century stressed that eugenics should be voluntary.

In the modern discussion of eugenics, there is great confusion over the term. It is used in many different ways. From the perspective of cognitive semantics, the concept of 'eugenics' exhibits prototype structure. There is a core of prototypical instances accepted by almost everyone, surrounded by more or less controversial nonprototypical instances radiating out at various distances. The dispute concerns how far the concept should be extended. Some prefer a narrow definition, others a wide one. Narrow definitions of eugenics often focus on state policy and coercion. Wide definitions, on the other hand, often focus on the intention to affect the kinds of people who will be born. Modern geneticists often prefer a narrow definition, because they want to dissociate present-day genetics from an abominable practice in the past and reassure the public of their good intentions. Critics of modern genetics, on the other hand, often use a wider definition to underline the continuity between old, morally unacceptable practices and the new genetics. Who is right? This is, of course, a mistaken way of phrasing the question. There is no 'true' meaning of the term. It can be used legitimately in many different ways. However, in this study I use a very broad definition:

> 'Eugenics' = def. an effort to affect the kind of person or persons who will be born.

Note that this stipulation presupposes intention. An effort is an intentional act. Other wide definitions might focus on consequences. This is the case with 'backdoor eugenics' (Duster, 1990). Here 'eugenics' denotes even unintended consequences of the availability of certain reproductive technologies. In my definition, however, eugenics concerns intentions, not with consequences, although eugenics in terms of intention might be an unintended effect of certain reproductive technologies or policies.

An advantage of this wide definition of eugenics is that one acquires an enlightening analytical tool that allows an integrated approach to the problem of the long term goals of reprogenetic medicine. The close relationship becomes obvious between initiatives taken by the government and health care professionals to affect reproduction in certain ways and attempts made by prospective parents to affect what kind of children they want. We see more clearly what is at stake than if we

[1] Quoted in Paul, 1994, p. 68.

reserve the term eugenics for initiatives taken by the government. A disadvantage of a wide definition such as the one suggested is that it includes even certain individual mating decisions. Another disadvantage is that it diverges too much from traditional terminology in medical history. In my opinion, however, the advantage outweighs the disadvantages.

Note also that this definition of eugenics comes close to the definition of 'genotypic prevention' put forward in Chapter 4: "an effort to prevent the birth of an individual with a particular clinical genotype." The difference is that genotypic prevention in this sense is a medical concept. According to my stipulation, it concerns 'clinical' genotypes only.[2] 'Eugenics' in the sense above, on the other hand, is not necessarily a medical concept. It may include also nonclinical genotypes, i.e. genotypes related to nondisease traits.

When trying to understand different kinds of eugenics in this intention-oriented sense, it is useful to distinguish between the following aspects:
(1) goals,
(2) agents,
(3) methods, and
(4) relation to science.
Let us take a closer look at each of these different aspects. They are particularly important for understanding modern versions of eugenics.

I will start with the goals of eugenics. All eugenic practices originate from an intention to affect what kinds of children will be born. However, a major distinction can be made between eugenic practices affecting populations or ethnic groups ('macro eugenics'), and those affecting only kinship groups, families or individual children ('micro eugenics') (cf. Lappé, 1995). Historically, eugenics was often stated in racial terms. It was identified with 'racial hygiene', i.e., improvement or impairment of qualities interpreted as linked to specific 'races'. The most typical example, of course, is Nazi eugenics. American and Scandinavian eugenics during the first three decades of the 20th century also commonly had this racial orientation, although from the end of the 1930s onwards it was more often stated in nonracial terms. The latter is also true of eugenics today. In some modern cases, the focus is on society at large. By improving the qualities of the population by reproduction control, the nation might become more competitive. An alternative view is to improve the human stock globally. It is a matter of humankind designing her own evolution. Other kinds of eugenics focus on public health. Here the perspective is medical, not political in the general sense (cf. Haker, 1993). Finally, we find a eugenics that is oriented towards individual children. Here, prospective parents make their own decisions about the kinds of children they wish to have.

---

[2] Of course, it would be possible to make a stipulation of 'genotypic prevention' that includes genotypes related to nondisease traits. However, the stipulation in Chapter 4 is proposed in the context of public health work, and therefore a reference to clinical genotypes is more appropriate (see p. 158).

It is important to ask about the agents of eugenic practices. Traditionally, eugenics is understood as a social policy with the government as the agent. By laws or other kinds of pressure the government directly steers reproduction. The reasons behind this policy may differ. As we have seen, it can be a broad social program aiming at improving public health or other qualities of the population and in this way making the nation more competitive. The reason might also be simply to cut health care costs. In national health care systems, professionals of reprogenetic medicine might function as mediators in such state-governed eugenic programs.

However, there might also emerge other forms of eugenics that depend on nonstate agents. Professionals of reprogenetic medicine working in private health care systems, may also strive for the goal of increased public health. Some of them might think that they should try to reduce the incidence of disease in the population by genetic methods. Moreover, there might emerge a 'market-driven eugenics' where insurance companies or biotechnology companies put pressure on prospective parents leading to eugenic action. In private health care systems, insurance companies might discourage prospective parents from giving birth to disabled children because of high costs. Families at genetic risk might not be accepted for insurance or be forced to pay very high premiums. Biotechnological companies might want to sell products, such as genetic test kits, which might end up in selective abortion. Although they probably have no eugenic objectives, the consequence might nevertheless be a kind of eugenics. This means, however, that the biotechnological companies themselves are only indirect agents for eugenics. Those who are the direct agents are the prospective parents who intentionally decide to use the technologies in order not to give birth to disabled children.

Saying this, I have already mentioned a third kind of nonstate agents, namely prospective parents (cf. Caulfield, Robertson, 1996, pp. 70-79). Parents might prefer children with certain traits and may use, for example, selective abortion on the basis of prenatal testing or, in the future, germline gene therapy to achieve that goal. One might talk about 'homemade eugenics' (Wright, 1990). Another term is 'consumer eugenics' (Juengst, 1995b). I suggest the term 'parental eugenics'. Parental eugenics might be the result of pressure from state or nonstate agents but it might also be a free choice.

Different methods can be used to achieve eugenic goals. It is common to distinguish between negative and positive eugenics. Negative eugenics concerns reduction of the frequency of 'negative' traits in the offspring. It can be described as a selection against undesirable properties. Positive eugenics, on the other hand, concerns increase of the frequency of 'positive' traits in the offspring. In this case there is a selection for certain traits. What traits that are selected for or against might differ. One might mean different things with 'desirable' and 'undesirable'. In this context, one might distinguish between disease-related eugenics and nondisease-related eugenics. Disease-related eugenics seeks to reduce the amount of dysfunctions or illnesses in offspring, and improve health. Eugenics that is not disease-related aims at reducing nondisease traits such as low intelligence, physical weakness or physical unattractiveness in offspring, and at increasing intelligence, physical strength or beauty. Of course, these two kinds of eugenics can be combined. Moreover, sometimes eugenicists talk about increasing or reducing 'traits', at other times about increasing or reducing 'genes' or 'genetic dispositions'. However, this is just a matter of speaking, since, according to an underlying assumption of genetic determinism, the traits are understood as being to a very large extent determined by the genes. Moreover, another distinction can be made between a concern for future persons and a concern for the impersonal gene pool. Sometimes eugenicists focus on future persons or generations, while at other times they focus on the gene pool.

When talking about methods one might also distinguish between nongenetic and genetic methods. Traditionally, eugenics was supposed to use nongenetic methods such as, on the one hand, sterilization, restrictions in immigration or, in the Nazi case, mass murder (the T4 program, and the Holocaust), to reduce undesirable traits, and, on the other hand, stimulation of 'good' parents through propaganda or economic advantages to increase desirable traits. Even certain modern versions of eugenics, such as the political program presented in *The Bell Curve* have this character. By reducing social welfare to certain ethnic groups in order to reduce the reproduction in these groups and imposing restrictions on immigration, Herrnstein and Murray argue that social problems will be solved and the nation will become more competitive (Herrnstein, Murray, 1994). In the modern medical context, however, gene technology rather than social control is intended to be used. More precisely, the methods might be selective abortion or preimplantation selection on the basis of genetic testing, or, in the future, germline gene therapy or genetic enhancement.

As illustrated by the *Declaration of Inuyama*, eugenics is often considered to imply coercion. Historical examples can be found in Nazi Germany, the US (Kevles, 1995 (1985)), and Scandinavia (Broberg, Roll-Hansen, 1996). However, some old eugenicists emphasized the voluntary nature of their proposals. One problem is, of course, where to draw the line between coercion and voluntariness. Coercion may be a matter of laws but it may also be a matter of, for example, economic or social pressure. How should one consider education? Should encouragement and discouragement of prospective parents be viewed as coercion? Moreover, eugenics may also be the unintended consequence of the stimulation of parental autonomy. Prospective parents may autonomously decide for themselves to give birth to a certain kind of a child without any influence from the state or nonstate agents.

Finally, we have the relation between eugenics and science. Eugenics can be based on either bad science or good science. By 'bad' and 'good' I here refer exclusively to the scientific quality in the narrow sense and exclude the ethical quality. Eugenics may be based on good science in this sense and at the same time be bad from an ethical point of view.

It is often stated that the old eugenics was based on bad science and that once we got rid of that bad science, eugenics ceased to be a serious alternative. However, as Diane Paul has shown, things were not as simple. It is true that the Hardy–Weinberg principle implies that if a trait is rare, most undesirable genes will be hidden in normal carriers as recessive genes, and, consequently, eugenic measures against those actually having the trait will be ineffective. Hundreds of generations will be needed before the trait disappears. According to Paul, however, many prominent geneticists continued to be eugenicists for a long time even after the articulation of the Hardy–Weinberg principle. One important reason for this was that 'bad' traits, such as 'feeblemindedness', were not considered to be rare at all. On the contrary, medicine and public charity were thought to keep the 'unfit' alive. They were even reproducing faster than the 'fit'. Some eugenic proposals undoubtedly were based on bad science. However, some were not. Paul stresses that what eventually made many geneticists take a stand against eugenics were social and ethical quandaries (Paul, Spencer, 1995; Paul, 1997).

Modern eugenic proposals may also be based on either bad or good science. Examples of the former could be eugenics based on genetic determinism. This determinism cannot, as a general view, be justified on the basis of present scientific findings (see pp. 107-110). However, it is also obvious that modern eugenics in a sense may be based on good science. If the goal is not to give birth to a child with a certain clinical phenotype and we do have some scientific knowledge of the penetrance of the genotype, then, of course, this kind of eugenics would be based on good science. Whether it is ethically acceptable is something completely different.

Can we learn anything from the history of eugenics? It exhibits a great diversity and, of course, some versions are more easily rejected than others. Today almost everyone would criticize the old state-governed eugenics using coercion to achieve its goals. Such clear examples of unacceptable eugenics should be opposed by the geneticists of our time as part of their social responsibility. However, there might be some other versions of a voluntary character that are unclear and that must be discussed.

The Human Genome Project and the 'New Eugenics'

Below I shall discuss more extensively two kinds of eugenics of special interest in modern democracies. Both are directly relevant to the question of the long term goals of reprogenetic medicine. The first version can be described as a 'voluntary public health eugenics', the second as a 'parental eugenics'. Both kinds of eugenics can be labeled 'new eugenics', since they in contrast to older eugenics use modern genetic technologies.

The 'new eugenics' was announced by the eminent molecular biologist Robert L. Sinsheimer in 1969. He saw the possibility of human genetic modification as "... a turning-point in the whole evolution of life. For the first time in all time a living creature understands its origin and can undertake to design its future." The old eugenics required a massive social programme carried out over many generations, the consent of the majority of the population, and social control. In contrast, the new eugenics could, at least in principle, be implemented on an individual basis, in one generation, and without social restrictions. The old eugenics would have required selective "... breeding of the fit, and a culling of the unfit." The new eugenics would allow, at least in principle, "... the conversion of all the unfit to the highest genetic level" (Sinsheimer, 1969, pp. 136, 144-145).

There are several things to note here. First, Sinsheimer is very speculative and takes genetic modification as a starting point. However, other possible methods of the new eugenics are selective abortion on the basis of genetic testing and screening, and to some extent preimplantation selection. In the foreseeable future, they are the only available methods. Sinsheimer does not mention these alternatives, but indirectly rules them out when he says that with the new eugenics "culling of the unfit" is unnecessary. Second, he stresses the new possibility of genetic enhancement, not only of gene therapy. Genetic methods should be used not only to treat diseases but to enhance human traits. And the enhancement seems to include not only disease-related traits but also nondisease-related ones. He even puts this possibility in an evolutionary perspective and indicates the possibility for what could be called 'autoevolution', i.e., the possibility of designing our own future evolution as a species (cf. Bayertz, 1994, pp. 70-73). Third, Sinsheimer understands the new eugenics as an individualist eugenics in contrast to the social breeding programs of the old eugenics. It is obviously, at least partly, a matter of a 'parental eugenics'. However, as mentioned above, another possible version of the new eugenics is 'voluntary public health eugenics'.

As a final comment, let me point out the key position of Sinsheimer both in relation to modern eugenics and the Human Genome Project. He not only announced the new eugenics in the late 1960s, but was also, alongside Charles DeLisi at the US Department of Energy (DOE), largely responsible for the initiative of starting the Human Genome Project. In 1985, Sinsheimer gathered a dozen leading molecular biologists to discuss the technical prospects of a human genome project (Kevles, 1992, pp. 18-19; Cook-Deegan, 1994, pp. 79-84, 92-106). It is interesting to see how Sinsheimer embodies both sequencing visions and eugenic visions. It is therefore tempting to conclude that for Sinsheimer there is an intimate connection between human genome research and eugenic applications. But, to be sure, this does not mean that his view is representative for geneticists in general.

4. GENOTYPIC PREVENTION VS. REPRODUCTIVE AUTONOMY: THREE PROTOTYPICAL VIEWS

Given this analysis of eugenics, let us now discuss more in detail the two possible goals of genotypic prevention and reproductive autonomy. What should the professional policy of reprogenetic medicine look like? Should the framework be a public health framework or a client-centered framework? Should the professional policy include genotypic prevention as a long term goal? How radical reproductive autonomy should prospective parents have? In order to find a realistic answer to these questions, responsible geneticists have to take into consideration certain contextual factors. In line with imaginative casuistry, it is vital that policy issues are reflected on with an eye on the social context. Responsible geneticists should envisage probable features of future democratic societies. In particular, they should take into account the disagreement and pluralism in society.

Moral Imagination: Probable Features of Future Democratic Societies

First, it is essential that responsible geneticists ask what features of a probable future society can be anticipated. They should not only reflect on what kind of society they want. They should do this in a realistic way. This means that they must take as a starting point certain features of our present democratic society that can be anticipated to be increasingly important in a probable future democratic society. These features can be seen as conditions that all serious thinking about genetics and the future of society has to handle. As far as I can see, the following features can be anticipated. I have already presented three of them in my discussion of gene patenting.

(1) *Pluralism.* We live in an increasingly pluralist society with many different and incompatible ethical views. Admittedly, there is some general consensus, but only at a very abstract level. There is agreement about ethical principles such as respect for autonomy, beneficence, nonmaleficence, justice, and so forth. However, when it comes to the interpretation, application, and weighing of these principles, there is no general consensus. They constitute a common heritage in our culture but there is nevertheless a plurality of uses. This can be seen in practical and emotionally loaded issues such as abortion, euthanasia, assisted reproduction, and environmental protection. The plurality of views is particularly obvious regarding gene technology. Moreover, we have no reason to believe that pluralism will decrease in the future. The lack of consensus is a fact that must be taken very seriously.

(2) *Scarcity of resources.* All public welfare systems and public health systems are characterized by limited resources. This means that when we discuss the application of genetics in such systems, economic aspects can never be neglected. Problems of priority-setting emerge. Moreover, the scarcity of resources is also a relevant fact in private health care. It is the basis of the market mechanisms. The buying and selling of genetic services obey the same laws as other market activities. It is obvious that scarcity of resources is something we have to live with for a foreseeable future, in the public sector as well as in the private one.

(3) *Commercialism.* Another characteristic of our present and future society is commercialism. No realistic policy for future applications of genetics can ignore

market pressures. Biotechnological and insurance companies have interests in reproductive as well as nonreproductive uses of gene technology.

(4) *Globalization.* Genetics research is an international enterprise, as is the development of gene technology. Moreover, certain kinds of research or development that are forbidden in one country may be permitted in another. The same is true of which testing techniques and therapies are clinically permitted. This has to be recognized by those who are developing policies for future applications of genetics.

Below, I shall focus primarily on ways of handling pluralism in society. This is particularly important as a background to the presentation below of three different prototypes for the view on genotypic prevention and reproductive autonomy as long term goals for reprogenetic medicine.

Moral Imagination: Democratic Authority vs. Individual Autonomy

How should the disagreement and pluralism in society be handled? This is a question of special relevance to the issue of reprogenetic policy. The genetics community itself has to handle pluralism one way or another. In addition, if the genetics community in its professional policy decides to assign the government a key role in reproduction, the problem is elevated to a higher level. Let us take a brief look at different views on how to handle pluralism at the governmental level.

However, this question of how to handle pluralism presumes that pluralism is a fact. This is not always recognized. Proponents of communitarianism often tend to tone down the pluralism in society. *Communitarianism* is here understood as the view that the development of society should be determined by the shared values that constitute our society.[3] But perhaps communitarians can accept pluralism as a fact? Some of them do. One example in the recent bioethics debate is Ezekiel Emanuel. He explicitly recognizes the pluralism of ethical views on health care issues. His solution to this problem is a *federation of political communities* dedicated to democratic deliberation. Each community would have their own health care facilities and establish their own health care policies (Emanuel, 1991, pp. 155-177). Emanuel's proposal is attractive but unrealistic. Such a federation of communities represents a turn in a direction quite opposite to the one we see in the US and Europe today with increasing centralization and bigger units. It could be an utopian ideal but hardly anything more.

Another communitarian alternative that accepts pluralism emphasizes the democratic process and maintains that whatever decisions come out of this process should hold. This means that *democratic authority* can be legitimately exercised even at the expense of minorities with conflicting views. We see this in all democratic countries. Examples are taxes, education and so forth. Why could it not

[3] Otherwise, communitarians often stay content with describing and analyzing society.

be accepted when it comes to reproduction and the application of genetics in reproduction? Some would argue that reproduction is too personal an issue to be determined by democratic authority. Others would maintain that the majority view should nevertheless rule. Yet others would argue that there at least should be certain democratically established restrictions on reproduction. Different views should be allowed, but only within certain limits. This third view is a serious and important alternative in the debate.

If we leave the communitarian framework and turn to an individualist one, there is another possibility of handling pluralism, namely *governmental neutrality*. The state accepts the plurality of views. This neutrality, however, can be justified in at least three different ways. According to the first view, neutrality is justified by appeal to *individual autonomy as the highest good*. According to the second view, neutrality is justified by appeal to certain values that are shared by all within the pluralist society. It is assumed that there is a *common core of values* in spite of disagreements regarding more detailed conceptions of good. This shared core includes, for instance, certain liberties. A third alternative is to defend neutrality as a *pragmatic strategy* of handling pluralism. Neutrality is viewed as a practical necessity. This pragmatic individualist approach does not rest on any assumptions about individual autonomy as the highest good or a common core of values. It accepts fundamental disagreement in society, and states that if we want to live together peacefully in a pluralist society, governmental neutrality is the only practical option.

I have focused here on governmental neutrality, but, of course, professional neutrality is also relevant. Even if the genetics community does not assign the government a key role in reproduction, it may decide to have professional neutrality as a professional policy.

It seems as if the main choice is between democratic authority and governmental/professional neutrality, or a certain mix of the two. However, I will not discuss this question in general terms. To be sure, this chapter is about the application of genetics in society, not about political philosophy.

Moral Imagination: Three Prototypical Views

With these contextual factors in mind, let us now turn to three different views regarding the professional policy of reprogenetic medicine on genotypic prevention and reproductive autonomy. They represent different interpretations, applications and rankings of common ethical principles. To some extent, they also represent different ways of handling pluralism in society, scarcity of resources, commercialism, and globalization. One view is individualist, the other two more communitarian. In my presentation and discussion of these views, I shall focus primarily on their interpretation of the ethical principles of autonomy and beneficence, and on their handling of pluralism. Moreover, these views represent different ways of handling the tension between a public health framework and a client-centered framework, and this will also be addressed.

With the terminology of imaginative casuistry, I call these views 'prototypical'. I do so in order to point out that they can be extended in various directions and be interpreted more or less radically. The prototypical views are:

(1) genotypic prevention as a key goal,
(2) radical reproductive autonomy, and
(3) reproductive autonomy with certain restrictions.

These views are closely related. The first stresses genotypic prevention as a key long term goal of reprogenetic medicine and is more or less restrictive regarding reproductive autonomy. The second and third prototypes include a rejection of genotypic prevention as a long term goal and stress the value of reproductive autonomy, more or less radically.

Let us take a closer look at each of these prototypical views.

Prototype One: Genotypic Prevention as a Key Goal

The first prototypical view is clearly communitarian. The framework is a public health framework, not a client-centered one. In order to be successful it probably requires a public health care system supported by the government. This means that the genetics community in its professional policy assigns government a key role in reproduction. The professionals of reprogenetic medicine and, ultimately, the government actively attempt to give reproduction in society a specific direction. A key long term goal of reproductive medicine is genotypic prevention, i.e. to prevent the birth of individuals with certain clinical genotypes. At a national level, the goal is to reduce the incidence of certain clinical genotypes in the population. In other words, reprogenetic medicine should try to determine what kind of people should be born.

If we focus on the role of government in reproduction and view the professionals of reprogenetic medicine as mediators in state-governed reproductive programs, we may make a distinction between 'strong' and 'weak' genotypic prevention. In the terminology of imaginative casuistry, these versions represent different extensions from core of the concept of genotypic prevention.

In 'strong' genotypic prevention, the goal of genotypic prevention is achieved by force, for example, by legal regulation. Note that if it is at all possible to realize the first prototype in a private health care system, strict legal regulation is probably needed. In 'weak' genotypic prevention, the goal is achieved by recommendations. These recommendations, mediated by the professionals of reprogenetic medicine, put prospective parents under pressure to reproduce in a certain way although they still have the possibility to refuse. The recommendations may be combined with certain economic advantages. There might also be mixed forms of governmental steering. Legal regulations regarding some aspects are combined with mere recommendations with respect to others.

Strong genotypic prevention was the dominating view in the eugenics in the first decades of the 20th century. A clear example is, of course, Nazi eugenics, which used laws and mass murder to achieve its goals. In the US, laws of sterilization and immigration were enacted (Kevles, 1995 (1985)).[4] In Scandinavia, coercive sterilizations were performed for several decades (Broberg, Roll-Hansen, 1996). Japan had eugenic laws until just recently.[5] However, some old eugenicists proposed weak versions, not least in the UK (Thom, Jennings, 1996). As we shall see below, most modern versions of genotypic prevention have a voluntary character. But there are modern proposals of strong genotypic prevention as well. An example is the political program presented in *The Bell Curve*. It can be described as an 'old eugenics in new clothes', because it shares the general social approach of the old eugenics but claims to based on modern science (Herrnstein, Murray, 1994).

Other modern versions of genotypic prevention are sometimes described as examples of a 'new eugenics' since they focus on the use of gene technology in reproduction rather than on broad social programs. More precisely, the methods to be used might be selective abortion or preimplantation selection on the basis of genetic testing, or, in the future, germline gene therapy or enhancement. In strong genotypic prevention of this new type, the government enacts laws governing the use of these genetically based methods in order to determine what kind of children should be born. In weak genotypic prevention of the new kind, on the other hand, the government puts prospective parents under pressure through education and dialogue to use certain genetic methods on certain diseases or traits. In the end, however, the parents are free to choose. The program is voluntary. Both kinds of governmental steering are based on democratic authority, i.e. they are established by majority decisions in the parliament (or its equivalent).

Different versions of this prototype differ with respect to the more specific content of legal regulation and governmental policy. They may differ about which genetic diseases are serious enough to justify genetic intervention. They may differ about the methods to be used. These differences may to some degree depend on whether the eugenic programs are part of a national health care system or whether they are to be implemented in private health care. Furthermore, it should be noted that the proponents of the first prototype in addition to genotypic prevention may also accept certain areas of reproductive autonomy and also certain restrictions of reproductive autonomy, i.e. certain elements from the second and third prototypes, respectively. I classify a particular view as belonging to the first category if it includes at least some elements of genotypic prevention. But it may include elements from the other two prototypes as well.

---

[4] See also, for example, Caulfield, Robertson, 1996, p. 59-70.

[5] For an analysis of the laws, see Kimura, 1991. The laws were changed in 1996.

I do not know of any explicit proponents of the new kind of strong genotypic prevention in the Western world but it is important to be clear about its possibility. Scarcity of resources makes it a nonnegligible alternative, although it might not be quite politically appropriate to proclaim it due to the dark past of its eugenic predecessors. China, however, is a non-Western country that tries to implement strong genotypic prevention (Mao, Wertz, 1997).

However, some people in the West explicitly argue for a weak genotypic prevention. I shall briefly present four different versions that have been put forward recently. The first is the proposal from the European Commission (1988) which was strongly criticized and later replaced. The second example is the beta-thalassaemia screening program in Cyprus in 1972-1983. Actually, I have already referred to these two cases earlier. The third example is a proposal by the philosopher Philip Kitcher. The fourth proposal is a suggestion by the philosopher LeRoy Walters.

Case: The European Commission Proposal from 1988

Let us start with the discussion preceding the creation of a human genome project in the European Community, which I have already mentioned twice in previous chapters. Let me repeat. In 1988 a proposal was presented by the European Commission with the title "Predictive Medicine: Human Genome Analysis." The program was called a 'health measure' and its aim was described in the following way: "Predictive Medicine seeks to protect individuals from the kinds of illnesses to which they are genetically most vulnerable and, where appropriate, to prevent the transmission of the genetic susceptibilities to the next generation."[6] The genome project was expected to make Europe more competitive by slowing down the rate of increase in health costs and by making its scientific and technological base stronger.

This proposal was criticized in a report by Benedikt Härlin, a member of the West German Green Party. He maintained that the proposal included eugenic ideas. As a result of this report, the European Commission put forward a new proposal that did not mention predictive medicine, simply called "Human Genome Analysis". It was formally adopted by the European Community Council of Ministers in 1990 to last for a three-year period.[7] After this period, the genome program became a part of the 'BIOMED' programs.

The argument from prevention presented in the original proposal is an example of a kind of justification of genome research that draws attention to the social value of applications. However, it was controversial and was eventually removed. It is interesting to see how the main argument for the program suddenly vanishes. The program is now presented as a purely scientific study, albeit with certain anticipated

[6] Commission of the European Communities, 1988; quoted in Kevles, 1994.

[7] See Commission of the European Communities, 1989; European Community, 1989; Official Journal of the European Communities, 1990; Kevles, 1994.

medical applications. Health care gives way to basic science. Does this imply that the concept of predictive and preventive medicine has completely disappeared? Probably not. It has been removed from the public agenda but it may still be on the hidden one. It must still be taken seriously. In particular, the strong economic incentives behind the original proposal make this assumption probable.

Case: Cyprus

We find a recent practical example of weak genotypic prevention in the beta-thalassaemia screening program in Cyprus in 1972-1983. As I have already mentioned, in a process of education and sensitization, marriages between couples in which both persons are carriers were discouraged. Later, antenatal diagnosis was introduced, making it possible for carrier couples to have children free of beta-thalassaemia. However, there was also considerable social pressure since the Church required a premarital certificate indicating that carrier testing was done. As a result, practically no new thalassaemic children have been born in Cyprus since the beginning of the 1980s.[8]

Kitcher and Walters

The next version of the first prototype has been put forward by Philip Kitcher, defending what he calls 'utopian eugenics'. He starts from a very broad definition of 'eugenics' and writes that "eugenic practice begins with an intention to affect the kinds of people who will be born" (Kitcher, 1996, p. 193). Moreover, he states that eugenics in this sense is inevitable. Once we leave our reproductive innocence behind, we understand that this is the case. However, he continues, there are many different forms of eugenics. The only possible form in modern democratic states is 'laissez-faire eugenics'. This is a form of eugenics based on reproductive freedom. If prospective parents want to affect the kinds of children to be born, they should be allowed to do so. However, Kitcher himself argues for what he describes as a particular version of laissez-faire eugenics, namely 'utopian eugenics'. The agent of this kind of eugenics is to some extent the government, since it is the government that actively takes a eugenic initiative in the first place. The government attempts to influence reproduction in society. In the end, however, prospective parents are free to make their own reproductive decisions. This is why Kitcher presents his utopian eugenics as a version of laissez-faire eugenics. The means used by the government to influence reproduction are education and public discussion. However, this education is not just a matter of giving neutral information. The content of education concerns "... what kinds of fetal characteristics would properly lead responsible people to terminate a pregnancy" (Kitcher, 1996, p. 203). When it comes to drawing

[8] For a discussion, see Hoedemaekers, ten Have, 1998.

a distinction between traits to promote and traits to avoid, Kitcher proposes that "... there should be some encouragement to draw the distinction in a particular way" (Kitcher, 1996, p. 203).

Kitcher argues that everyone, apart from strict antiabortionists, accepts that when it comes to certain very serious genetic diseases the quality of life is so low that selective abortion is justified. It should therefore be recommended by the government in an educating dialogue with prospective parents. In addition, Kitcher rejects selective abortion concerning minor afflictions, disease susceptibilities, and nondisease traits such as sex. On the other hand, he accepts full freedom of choice in the gray zone between very serious diseases and less serious conditions. In this grey zone, the government should give no specific recommendations (Kitcher, 1996, pp. 206, 299-300).

As far as I can discern, Kitcher's eugenics is called 'utopian' because all should be given access to these genetic facilities and because the possible success of the program is the contingent outcome of this educational initiative.

Kitcher does not explicitly present his proposal as a public health program but it can reasonably be interpreted in this way. He talks, for instance, about "... a program to lower the incidence of genetic disabilities" (Kitcher, 1996, p. 85). Relevant is also his suggestion to eliminate medical insurance entirely and develop "... a system in which care is available to all citizens and funded through progressive taxation" (Kitcher, 1996, pp. 135-136). At the foundation of Kitcher's eugenic proposal, there is a vision of a national health care system.

The main method to be used in this public health program is selective abortion based on prenatal testing. But Kitcher seems to think that when germline gene therapy is technically available it could also be used. He is even willing to contemplate the genetic enhancement of, for example, our ability to resist infectious disease (Kitcher, 1996, pp. 123-125). Moreover, Kitcher's eugenic public health program is voluntary. It is obvious, however, that governmental 'encouragement' puts some pressure on prospective parents. This means that the program is not 'truly voluntary', to use the words of the *Declaration of Inuyama*. The aim of the government is not to the increase the procreative liberty of the citizens but to direct their reproduction in a certain way so that certain kinds of children are not born.

A key question is, of course, where to draw the distinction between desirable and undesirable traits. Kitcher uses quality of life as a criterion, not disease or health. He distinguishes between three different 'dimensions' of quality of life: (1) a sense of what is significant, (2) satisfaction of central desires, and (3) the balance of pleasure and pain. On the basis of these criteria, Kitcher finds some clear cases for genetic intervention constituting two different extremes on a continuum. At one extreme, he finds cases with no chance for self-determination, such as early-onset neurodegenerative disorders, and cases with low quality of life combined with large impact on others, such as fetuses with a degenerative muscular disease and where the parents already have problems with existing children. According to Kitcher's judgment, everyone, except those who are firmly against all kinds of abortion, can agree about these 'clear' cases. At the other extreme, he finds cases with somewhat greater than 50% risk that a heart disease will strike in middle age. Here, consensus is clearly against abortion. Between these cases are the hard ones. An example is cystic fibrosis of the common form. In this case the seriousness of the phenotypic expression may vary a lot; it is impossible to predict how it will be; and there is also the possibility of new therapies. Different prospective parents may come to different conclusions (Kitcher, 1996, pp. 285-300).

I mentioned above that Kitcher thinks that the main method for reduction of the incidence of serious genetic diseases is selective abortion, and only very indirectly indicates the possibility of using germline gene therapy. LeRoy Walters (together with Julian Gage Palmer), on the other hand, explicitly stresses germline gene therapy as an important method in a future public health program. In contrast to Kitcher, Walters does not want to use the term 'eugenics' since it is so tainted by the old eugenics. It has too negative a connotation. But Walters accepts the idea of a morally justifiable eugenics, namely one that is strictly voluntary and based on reliable genetic technologies. He describes his 'public-health program' as "a voluntary program to reduce the incidence of genetic disease through germline genetic intervention" (Walters, Palmer, 1997, p. 88). He means that germline gene therapy on serious diseases can be an important public health measure in the future. In this way, selective abortion can be avoided. This is also more in line with the general goal of health care to cure rather than to terminate life.

Summary

The original European Commission proposal, the Cyprus example, Kitcher's proposal, and Walters' proposal represent different voluntary versions of the first prototype. The agent of this kind of 'voluntary public health eugenics' is the government using the professionals of genetic medicine as mediators. The central idea is genotypic prevention as a public health measure aimed at the reduction in the incidence of certain serious genetic diseases in the population. Possible methods to achieve this goal are selective abortion on the basis of prenatal testing and screening, preimplantation selection, and germline gene therapy. Prenatal testing and selective abortion as well as–at least to some extent–preimplantation selection are available at present, while germline gene therapy is not and will probably not be for a long time. The government puts prospective parents under some pressure to use such methods on serious diseases. It does this by encouraging them through education and dialogue. This means that genetic counseling is to some extent directive. In the end, however, the parents are free to choose. The public health program is voluntary, based on a respect for reproductive autonomy.

Arguments in Favor

The main argument in favor of genotypic prevention, whether strong or weak, is improved public health and reduction of the amount of dysfunctions or illnesses in future generations. On the basis of genetic testing and screening, selective abortion or preimplantation selection may be used to avoid giving birth to persons suffering from genetic diseases. Moreover, by germline gene therapy, if it becomes available sometime in the future, genetic diseases may be eliminated for good in a particular family line. Fundamentally, this argument is based on the principle of beneficence. It is a matter of preventing suffering in future generations by avoiding that certain kinds of children are born. The principle of beneficence ranks higher than the principle of autonomy but to a varying degree depending on whether we talk about weak or strong governmental steering. In the former case, the autonomy of the prospective parents is respected to some extent, in the latter democratic authority outweighs it. In both cases, pluralism in ethical views on reproduction is handled by majority decisions.

Another argument concentrates on fair distribution of limited resources. The argument is relevant only within a national health care system. Initially it concerns only selective abortion based on prenatal testing, not preimplantation selection or germline gene therapy. Even when preimplantation selection becomes generally available and germline gene therapy is developed, these measures will probably be very expensive. According to this argument, genotypic prevention through a program for selective abortion of fetuses with certain serious diseases is a matter of economic realism and justice. In this way it will be ensured that there will be enough money left for other groups of patients with more positive prognoses.

Each of these arguments, or both, can be supported by a communitarian argument stressing the importance of 'responsible reproduction'. Society may demand certain things of its citizens. Sometimes the interests of society have priority over the interests of individuals. We accept democratically decided coercion in many cases. Democratic authority, not full individual autonomy, is the key value of modern democratic societies. Society has certain legitimate interests in reproduction. If genotypic prevention can prevent the suffering of future generations and reduce the costs for health care, society may put prospective parents under pressure in order to implement it. The most efficient way, however, does not seem to be coercive laws but education and encouragement of prospective parents. Reproduction is a very private thing to many people and a strong resistance to eugenic laws might emerge.

These arguments may be supplemented by an argument making a comparison between the effects of 'voluntary public health eugenics' view and the 'radical reproductive autonomy' view. The latter might lead to eugenics anyhow, namely to a 'parental eugenics'. This kind of eugenics may have unpredictable consequences. It is much safer if the government and the professionals of reprogenetic medicine influence the direction of eugenics by encouraging prospective parents to make certain reproductive choices rather than others.

Prototype Two: Radical Reproductive Autonomy

The second prototype is 'radical reproductive autonomy'. It represents a radical individualist and client-centered framework. In order to understand this alternative, it is important to start with a historical perspective. As mentioned above, respect for reproductive autonomy is a well-established principle in clinical genetics that is expressed in the practice of nondirective genetic counseling. This practice is traditionally thought to be an antidote to eugenics. The second prototype takes the respect for reproductive autonomy to its extreme. Pluralism in ethical views on reproduction is handled with a policy of strict professional and governmental neutrality.

The principle of reproductive autonomy is commonly thought to be applied within the framework of private health care, but it may also be applied within a public health care system. Moreover, reproductive autonomy is commonly understood as a negative right in the sense that other people have a duty not to interfere with the procreative choices of prospective parents. However, it is quite possible to argue that respect for reproductive autonomy also has a positive side. The government has a responsibility to provide at least some means to ensure that reproductive autonomy not merely becomes a privilege for the rich. The goal is to strengthen and increase the reproductive autonomy even of ordinary people. This policy can supported by the principle of justice.

It is easy to imagine that the implementation of this model might lead to far-reaching parental eugenics. 'Parental eugenics' can be defined, as was done above, as a kind of eugenics performed by prospective parents with the intention of affecting the kind of children they will have. It is a matter of genotypic prevention in the sense of a measure of prospective parents aiming at avoiding to give birth to children with certain genotypes. In this kind of eugenics, prospective parents, not the government or the professionals of reprogenetic medicine, are allowed to use genetic methods to achieve eugenic goals. The second prototype does not include genotypic prevention as a goal of reprogenetic medicine. It only accepts genotypic prevention as a possible unintended consequence of the implementation of respect for the reproductive autonomy of prospective parents. Note that this kind of eugenics does not seem to be understood as eugenics in the *Declaration of Inuyama*. In this *Declaration*, "truly voluntary programmes" are contrasted to eugenics.

Moreover, this prototype allows selective abortion not only of fetuses with serious genetic diseases but also of those with nonserious diseases, and certain nondisease characteristics such as sex (in absence of sex-linked genetic risks). It also allows preimplantation selection with respect to various traits and afflictions. Moreover, this model opens up the possibility for using somatic cell gene therapy on fetuses, germline gene therapy, and enhancement, although these technologies are not available at present.

I would also like to correct two common misunderstandings of the second prototype. First, it is often maintained that it gives market forces free scope. This is not necessarily so. A reasonable interpretation of 'radical reproductive autonomy' demands protection not only from pressures from the government but also from biotechnology companies and insurance companies. The obligation of providing such protection belongs to the responsibility of the government and the professionals of reprogenetic medicine. Second, it might be thought that the second prototype with necessity starts from an atomistic view of the ontological relation between the individual and society. However, it is quite possible to start from a modified social holism. In this case, the individual to be sure is viewed as a product of social influences but also as a unique combination of such influences. In this way, individuality is preserved within the conception of a social network (see pp. 5-7, 57-58).

Case: Northern India

A practical example of the consequences of radical reproductive autonomy can be found in Northern India. In this region, prenatal testing followed by selective abortion of female fetuses is not uncommon. However, it is important to look at this sex selection in a historical perspective. Female infanticide has been practiced here for centuries. Modern medical technology has only changed the methods (Kumar, 1985).

Robertson

An important proponent of the 'radical reproductive autonomy' view is John Robertson (Robertson, 1994). He understands 'radical reproductive autonomy' within the framework of private health care and stresses that full procreative liberty, or radical reproductive autonomy, is an important right.

However, it is a right only in the 'negative' sense "that a person violates no moral duty in making a procreative choice, and that other persons have a duty not to interfere with that choice" (Robertson, 1994, p. 23). It is not a 'positive' right to have the government provide the necessary resources.

Robertson's argument is that "procreative liberty deserves presumptive respect because of its central importance to individual meaning, dignity, and identity" (Robertson, 1994, p. 16).

Arguments in Favor

The main argument for the second prototype is a radical interpretation of the principle of respect for reproductive autonomy. According to this interpretation, respect for reproductive autonomy is not only a principle for the implementation of the goals of reprogenetic medicine but a goal in itself. The professionals of reprogenetic medicine and the government should make efforts to increase the autonomy of prospective parents. They should not try to influence them in a certain direction. Such governmental pressure would be a violation of reproductive autonomy. What prospective parents need is sufficient information to make up their own minds.

But, one might ask, why should we lay particular stress upon the principle of autonomy precisely when it comes to reproduction? In many other areas of health care, physicians recommend certain alternatives of action, for example a particular treatment, even if the patients in the end have the right to say 'no'; why should not clinical geneticists, obstetricians, and genetic counselors give recommendations regarding reproduction? Why should we prefer a radical interpretation of the principle of respect for autonomy compared to a more moderate one?

The high appreciation of reproductive autonomy can be supported by many different arguments. These arguments can be used by proponents of different ethical approaches: rights-based, utilitarian, casuistic, etc.

One important reason has already been mentioned above in the brief presentation of Robertson's view. Reproductive autonomy should be respected because of its central importance to individual meaning, dignity, and identity. Procreation is extremely valuable to the individual and heavily existentially loaded, and should therefore be left to the individual to decide about. It is valuable because it implies participation in the creation of a person, because it is an expression of mutual love and sexual intimacy, and because it is a link to future persons. Moreover, the experiences of pregnancy, childbirth and child rearing are very precious (cf. Strong, 1997, p. 79).

Another reason for a high appreciation of reproductive autonomy is that there is a pluralism and lack of consensus with regard to the ethical aspects of reproduction. In ordinary health care, there is commonly a reasonable consensus among physicians about what kinds of action are appropriate. However, there is no consensus about, for instance, where to draw the line for acceptable selective abortion and genetic intervention. Admittedly, there are some clear cases on which everyone but strict anti-abortionists agree. But there are also many unclear cases with no general consensus. In the latter cases, the professionals of reprogenetic medicine do not know which are the best decisions, because it might be impossible to predict on the basis of genetic testing how severe a disease will be, and because there are many different moral views in our modern society. With this unpredictability and pluralism in mind, it is extremely important that prospective parents be allowed to decide for themselves without pressures of any kind. If the government tries to influence reproduction in a certain direction, the personal, deeply held views of many people are violated.

A third argument is that it is extremely dangerous if the government makes the 'wrong' decisions in its steering of reproduction. This is much more dangerous than if individual parents make them. The social consequences of mistakes made by the government are of a much bigger magnitude. Clear examples are the discrimination and stigmatization resulting from the eugenic policies during the first half of this century. The history of eugenics makes not only mandatory but even voluntary governmental steering of reproduction an unattractive alternative.

One further reason for accepting a policy of radical reproductive autonomy is that prospective parents themselves are clients, not patients. They come to the clinical geneticist not to receive treatment, but to get the information they need to make a well-considered reproductive choice. Directive genetic counseling would be almost an insult to them.

A final, and more pragmatic, reason is that a policy implying that the government and the professionals of reprogenetic medicine should try to direct reproduction in a certain way would probably be unsuccessful, because reproduction is commonly perceived as one of the most private things in our culture and because many people consequently would perceive such initiatives as a violation of their privacy and would react strongly against it.

Prototype Three: Reproductive Autonomy with Certain Restrictions

In the third prototype, reproductive autonomy is still a very important principle. The arguments in favor of a high appreciation of reproductive autonomy are to a large extent accepted. However, certain restrictions are placed on reproductive autonomy. Prospective parents are allowed only within certain limits to decide for themselves what kind of children they want to give birth to. Pluralism in ethical views is handled by professional and governmental neutrality, but only within certain boundaries. The government does not try to directly steer reproduction; it only provides certain restrictions. These boundaries are based on democratic authority. This prototype can therefore be classified as a communitarian one, as can the first prototype. Moreover, this view, like the previous one, can be combined with active efforts by the government, based on the principle of justice, to see to it that certain genetic facilities are in place for those who want to make use of them.

Within the democratically determined boundaries, prospective parents are allowed to decide for themselves what kind of children they will give birth to. This means that parental eugenics is accepted as a possible outcome of parental autonomy, but only to a certain extent. Certain restrictions are placed on parental eugenics. Note, as is the case with the second prototype, the third prototype does not include genotypic prevention as a goal of reprogenetic medicine. It only accepts genotypic prevention as a possible unintended consequence of the implementation of respect for the reproductive autonomy of prospective parents. However, in contrast to the second prototype, it imposes certain restrictions on this possible parental eugenics.

There are many different versions of the third prototype, and they differ in at least two fundamental respects. First, the restrictions may differ as regards normative status. They may be legal restrictions or only recommendations. In the former case, we can talk about 'strong' restrictions, in the latter about 'weak' restrictions. This is parallel to the strong and weak versions of the first prototype. Second, the subjects of restriction may vary. The restrictions might concern which characteristics should be allowed to prenatally test for: very severe diseases, severe diseases, less severe diseases, minor afflictions, disease susceptibilities, late-onset disorders, or nondisease traits such as sex (in the absence of sex-linked genetic risks) or intelligence (if possible in the future). They might also concern which interventions on the basis of test results should be allowed to be carried out: selective abortion, preimplantation selection, somatic cell gene therapy on fetuses, germline gene therapy, or genetic enhancement.

Council of Europe: Convention on Human Rights and Biomedicine 1996

Let me give an example of an official policy document that is in line with the third prototype, namely the *Convention on Human Rights and Biomedicine* issued by the Council of Europe in 1996. In Article 14 there is a statement against sex selection (Council of Europe, 1996):

> The use of techniques of medically assisted procreation shall not be allowed for the purpose of choosing a future child's sex, except where serious sex-related disease is to be avoided.

Moreover, in Article 13 appears a disclaiming of nondisease genetic intervention and also implicitly of germline gene therapy (Council of Europe, 1996):

> An intervention seeking to modify the human genome may only be undertaken for preventive, diagnostic or therapeutic purposes and only if its aim is not to introduce any modification in the genome of any descendents.

Both articles place clear restrictions on the reproductive autonomy of prospective parents, and also, of course, on the freedom of choice of the professionals of reprogenetic medicine.

Murray and Strong

An important proponent of the third prototype is Thomas H. Murray. In his book *The Worth of a Child* he devotes a chapter to the question of prenatal testing and nondirective genetic counseling. He maintains that nondirectiveness is a way for genetic counselors to "parry accusations of eugenics." They can "appeal to our culture's infatuation with autonomy" (Murray, 1996, p. 122). However, according to Murray, "an ethic of nondirectiveness is wholly unprepared for the coming avalanche of genetic information" (Murray, 1996, p. 128). It offers no help in dealing with adults who want to use genetic information to choose sex, hair, eye color, height, tendency toward obesity, etc. With respect to very serious and serious genetic diseases, Murray accepts the reproductive autonomy of prospective parents. They should be permitted to decide for themselves. However, concerning sex, traits and afflictions that are not substantial burdens on parents and children, Murray argues that we should not offer to provide such information prenatally. Moreover, with Adrienne Asch, he stresses the importance of giving prospective parents information about what it is like to live with a disability and what it is like to raise a child with a disability (Murray, 1996, p. 139; Asch, 1989). A certain directiveness in genetic counseling is important in order to fight 'perfectibilism', i.e. the quest for the perfect child. Perfectibilism is, according to Murray, incoherent with the meaning of parenthood. It is one thing to want a child, a completely different thing to want a particular child, and even more so, a 'perfect' child (Murray, 1996, pp. 118-119, 130-137).

Another proponent is Carson Strong. In his recent book *Ethics in Reproductive and Perinatal Medicine*, he argues for a version of the third prototype that draws the line between disease and nondisease traits. Strong applies this view to selective abortion and preimplantation selection but does not discuss its application to somatic cell gene therapy on fetuses, germline gene therapy, or enhancement (Strong, 1997, pp. 135-157).

Arguments in Favor

As mentioned above, the arguments in favor of a high appreciation of reproductive autonomy put forward by the proponents of the second prototype are to a large extent accepted also by the proponents of the third prototype. However, certain restrictions are placed on reproductive autonomy. Many different reasons have been offered in favor of such restrictions, and, of course, they differ depending on exactly what restrictions are suggested. However, some reasons recur. This makes them, on the other hand, rather difficult to assess. They can be put forward in favor of radical as well as less radical standpoints. The reasons are of two kinds. The first kind states that certain genetic interventions are morally wrong in themselves. For instance, they are said to violate human dignity or human rights. The second kind of reasons stresses that certain interventions might have bad consequences. They focus on social risks, such as discrimination and stigmatization, or other kinds of risks.

As an example, let us concentrate more specifically on one kind of intervention that might be subject to restrictions according to the third prototype, namely selective abortion. I shall briefly present two different views on this kind of intervention and some arguments in favor of them.

According to one position, selective abortion should only be allowed regarding very severe and severe diseases. This view is based on the premise that abortion should be avoided as much as possible because of the worth or dignity of the fetus. Only very severe and severe diseases imply a low enough quality of life or serious enough suffering to justify selective abortion. Moreover, if the government allows selective abortion of fetuses with less severe diseases, this might lead to discrimination and stigmatization of people living with these diseases. Selective abortion of fetuses with such diseases should, therefore, not be allowed. Reproductive autonomy is important but it can only be accepted within these strict moral boundaries.

Another position allows selective abortion not only on fetuses with very severe or severe diseases but also on those with less serious diseases, minor afflictions, disease susceptibilities, or late-onset disorders. However, it does not accept selective abortion regarding nondisease traits such as sex (in absence of sex-linked disease). Within the boundaries of disease, this position stresses the importance of respect for reproductive autonomy. Conversely, this implies the criticism of the other position for not taking reproductive autonomy seriously enough. A reason for not offering parents the opportunity to make reproductive choices regarding nondisease traits, including sex (in absence of sex-linked disease), is that such traits have nothing to do with the health of the fetus or subsequent person. Health care professionals have no obligation to give information prenatally about such traits and make interventions on the basis of such information. A reason for allowing selective abortion not only on fetuses with severe diseases but also on fetuses with less severe diseases, minor afflictions, disease susceptibilities, or late-onset disorders, is lack of consensus about where to draw the line. The distinction between disease traits and nondisease traits seems less fuzzy and includes less subjective value judgments than the distinction between serious and nonserious disease, even if it is not completely strict. With regard to the risk for discrimination and stigmatization of persons living with less severe diseases or afflictions, proponents of this position argue that such social consequences can be and should be counteracted. It is quite possible to create a good social environment for such people even if the government allows selective abortion on fetuses with such diseases or afflictions.

Moral Imagination: The Character of the Dispute

How should these different prototypes be assessed? Is any version of the prototypes to be preferred? In reprogenetic medicine, how should the pluralism of ethical standpoints, scarcity of resources, commercialism, and globalization be handled? How should the central ethical principles be interpreted, applied and balanced in this context? What is to be preferred, an individualist approach or a more communitarian one? What role should health professionals and the government play in reproduction? Responsible geneticists need to use their moral imagination to answer these questions.

The arguments for and against each of these prototypes can be described as centered around commonly recognized ethical principles. From the perspective of imaginative casuistry, the disagreement among proponents of these views depends on different interpretations of these ethical principles, on different views of their scope of application, and on different ways of balancing them against each other when they conflict. Thus, communitarians and others are right when they stress that there is general agreement in society on important ethical principles, but critics are right when they stress the disagreement about their interpretation and application. Most people accept prima facie such principles as respect for autonomy, beneficence, nonmaleficence, justice, integrity, and equal human value, but the principles can be applied to the question of the long term goals of reprogenetic medicine in many different ways. Let us as an illustration compare how two of these principles are understood and used within the three prototypes, namely the principle of respect for autonomy and the principle of beneficence.

All prototypes and all versions of them, except strong genotypic prevention, stress the principle of respect for autonomy. But there are some important differences. Weak genotypic prevention gives the principle of autonomy a moderate interpretation. It accepts that the professionals of reprogenetic medicine and the government put prospective parents under pressure by encouraging them to avoid giving birth to children with certain clinical genotypes, although the parents in the end have the right to say 'no' to these recommendations. The third prototype also interprets the principle of autonomy in a moderate manner. The professionals of reprogenetic medicine and the government try to prevent prospective parents from making certain kinds reproductive choices, for example by discouraging sex selection. The second prototype, on the other hand, interprets the principle in a radical way. In contrast to the weak versions of the first and third prototype, it aims at strengthening the reproductive autonomy of prospective parents. Prospective parents should be encouraged to choose freely, to make up their own minds, even if this might have far-reaching consequences in terms of parental eugenics. Moreover, as stressed above, the prototypes can be described in terms of political framework. The second prototype is clearly individualist, while the first and the third prototypes are communitarian.

The principle of beneficence may also appear in more or less radical applications. In a moderate application, the scope of the principle is rather narrow. The principle covers only individuals of the present generation. In a radical application, it covers also future generations. The first model makes a radical application. It finds the principle of beneficence so important that the government in order to improve public health in future generations should encourage prospective parents to prevent the birth of children with certain severe genetic diseases. Prospective parents should not be encouraged to choose as they wish in cases of severe genetic diseases but should be put under pressure, legal or otherwise, to choose in a certain way. Only children who will not suffer from severe genetic diseases should be allowed to be born. However, in contrast to this, the second and third prototypes stress that the professionals of reprogenetic medicine and the government should refrain from putting prospective parents under pressure not to give birth to persons who would suffer from serious genetic diseases. They make a moderate application of the principle of beneficence. Prevention of the birth of children is acceptable as a reproductive choice of prospective parents but not as a public health measure of the government.

In conclusion, we see that the three prototypical views are based on different interpretations, applications, and rankings of commonly accepted ethical principles. The choices between these are not easy. Nor is it easy to choose between the three prototypes as such. The choices are fundamentally a matter of moral imagination. What should a professional policy on these issues look like? It is obvious that a serious dialogue within the genetics community on these views is essential. In order to reach a well-considered decision, however, a dialogue between geneticists, other scientists and scholars, politicians, industrialists and the general public is also needed. Let me end this chapter by presenting a proposal of my own as a contribution to such a dialogue.

5. PROPOSALS REGARDING THE LONG TERM GOALS OF REPROGENETIC MEDICINE

My proposals regarding the long term goals of reprogenetic medicine include a disease-centered version of the third prototypical view. Reprogenetic medicine should not have the goal of genotypic prevention, other than perhaps by way of extreme exception. Reproductive autonomy is, generally speaking, too important. Reproductive autonomy should the point of departure. In discussing difficult, nonprototypical cases, the burden of proof should lie on those who suggest genotypic prevention. Genetics should primarily serve the interests of those directly involved in reproduction, i.e. the fetus and the parents, not future generations or the impersonal gene pool. This high appreciation of reproductive autonomy means that the approach is fundamentally client-centered, not public health-centered. When there is a tension between public health objectives and client objectives, the latter generally outweigh the former.

The reasons for this proposal have actually been presented already. Procreation is very important to individual meaning, dignity, and identity. Issues related to procreation are therefore often heavily existentially loaded, and are characterized by a lack of consensus. The best way to handle this kind of pluralism is professional and governmental neutrality, not to force majority decisions on minorities.

However, we accept democratically established restrictions on individual autonomy in many other aspects of life, and it seems reasonable to accept certain restrictions in reproduction as well. The government and professionals of reprogenetic medicine cannot remain completely neutral. There have to be certain restrictions on the parental use of gene technology in reproduction. There are views that deviate too much from the majority's value commitments. The government and the professionals of reprogenetic medicine have to place certain restrictions on procreative liberty, if only for the pragmatic reason of preserving harmony in society. The restrictions should be a matter of open discussion within society.

Regarding such restrictions on reproductive autonomy, I suggest that they should be disease-centered. In this I come close to Strong. However, I apply this view not only to selective abortion and preimplantation selection, as does Strong, but also to somatic cell gene therapy on the fetus, germline gene therapy, and genetic enhancement (see p. 245). Within these boundaries, prospective parents should be allowed to choose autonomously whether to use certain genetic technologies. By implication, this means that there should be no differences in principle between the uses of different genetic technologies with regard to the seriousness of the diseases approached. For instance, the use of preimplantation genetic diagnosis should not be restricted to only very serious diseases while other prenatal genetic testing is allowed to also address less serious diseases. In the short perspective, this may be justified due to various contextual reasons, but not as a long term goal.

My general reasons for these restrictions are those already presented in the section about the third prototype. To draw the line between disease and nondisease characteristics seems more reasonable than to do so between serious and less serious diseases, or between some nondisease traits and others. Reprogenetic medicine should keep to the domain of medical conditions, and the distinction between disease and nondisease traits seems much less fuzzy and includes less subjective judgments than the other two distinctions, although it is not completely strict. Moreover, within the domain of disease traits, the unclear border between serious and nonserious diseases is a further reason to leave the choice to the prospective parents themselves.

Demarcation of Disease

This argument makes the demarcation between disease and nondisease crucial. Is the distinction between disease and nondisease actually less fuzzy and less subjective than the distinctions between serious and nonserious disease and between some nondisease traits and others? Note that it is not necessary in order to make the argument work that the definition of disease is absolutely objective or the border between disease and nondisease absolutely strict. Since the distinction between serious and less serious diseases and the distinction between some nondisease traits and others are very subjective and fuzzy, it is sufficient to show that the distinction between disease and nondisease is less subjective and less fuzzy.

Many different definitions of disease have been proposed. Some have been objectivist (e.g. Boorse, 1977), others subjectivist (e.g. Engelhardt, 1986). A third set constitutes a middle course (e.g. Clouser et al., 1981; Nordenfelt, 1995 (1987)). From the perspective of cognitive semantics, it is obvious that we can interpret the concept of disease in many different ways. From this point of view, it is also obvious that in defining disease different metaphors are used. Examples are 'disease is malfunction' (e.g., "Her heart does not *circulate* the blood properly"), 'disease is abnormality' (e.g., "His liver had an *abnormal* appearance"), 'disease is attack' (e.g., "She is under *attack* from viruses", "The cancer *defeated* him"), 'disease is distress' (e.g., "She *suffered* from cancer"), and 'disease is disability' (e.g., "His disease *hinders him from doing* what he used to"). In many definitions different metaphors are combined.

The concept of disease clearly includes evaluative and social dimensions. Take the case of 'hysteria'. At the end of the 19th century, it was viewed by many psychiatrists as a disease. Now, however, no one would classify the behavior as such. It is clear that what is considered a disease to some extent may vary over time and between different societies. Actually, this is what could be expected from the point of view of cognitive semantics. However, cognitive semantics does not imply that there can be no objective dimension at all. I shall try to show that although the concept of disease includes evaluative and social dimensions, there is nevertheless also an objective dimension. This can be seen even in the case of hysteria. Further research has shown that it is not properly classified as a disease since it does not fulfill certain scientific conditions.

The objective dimension of disease can be interpreted in at least two ways, and these two interpretations can be combined. One interpretation focuses on biological abnormality. The other is made in terms of an attack by ontological entities such as genes, germs, or environmental substances, through a pathological process. It is sufficient for my argument if one of these interpretations is plausible. The argument becomes even stronger if this holds for both of them. Note that I will not propose a definition of disease.

Let us start with biological abnormality. Robert Wachbroit has shown that biologists talk about biological normality/abnormality as distinct from statistical normality/abnormality and evaluative normality/abnormality (Wachbroit, 1994a). According to Wachbroit, 'normal' is used in explanations in biology much as 'pure' is used in explanations in chemistry. Physiologists talk about a 'normal' heart like chemists talk about 'pure' water. He stresses that "[j]ust as chemical purity is distinct from any moral or cultural understanding, so biological normality is distinct from any moral or cultural understanding one might have about normality" (Wachbroit, 1994b). In other words, biological normality is distinct from evaluative normality. According to this latter concept, being normal is being so according to the values of the majority of the population. And evaluative abnormality is a deviation according to the values of the majority of the population. Disease is something that is devalued by most people. By implication, what counts as disease may vary with changes of the values of the population. This does not seem reasonable from a biological point of view. In order to be able to account for the possibility of having a disease but being normal according to the values of the majority, we have to talk about a biological abnormality in distinction from evaluative abnormality.

Wachbroit also stresses that biological normality is distinct from statistical normality. Statistics may be important evidence for determining normality, but they are not part of the meaning or definition of normality (Wachbroit, 1994b). Statistical normality is what characterizes the average of the population or a standard portion of it, for instance, the majority. Statistical abnormality is a deviation from average or standard characteristics. Disease is something that the average person or most people do not have. This means that what counts as disease may vary with changes in the population. This does not seem reasonable from a biological point of view, either. The state of the normal heart does not need to be the state of the average heart or most hearts. In order to be able to account for the possibility of having a disease but being statistically normal, we have to talk about a biological abnormality in distinction from statistical abnormality.

Moreover, the concepts of biological normality/abnormality are often systematically linked to the concepts of biological function/malfunction. The exact relation is a matter of some dispute. Some argue that biological normality should be understood in terms of biological functioning, others that biological functioning should be understood in terms of biological normality (Wachbroit, 1994a). However, it is not necessary for my argument to settle this question. The important thing is that physiology determines whether something, for example a heart, is functioning or malfunctioning, and whether it is in a normal state or an abnormal. Biological function/malfunction and biological normality/abnormality is matter of physiological theory and physiological judgment.

That biological abnormality is part of the objective dimension of the concept of disease means that it is a necessary condition for disease. If someone manifests disease, there must exist a biological abnormality. This does not mean that the individual must subjectively experience the disease. Neither does it mean that the biological abnormality can actually be detected. Further research might be needed.

The second interpretation of the objective dimension of disease is ontological. To be sure, such an ontological understanding is somewhat old-fashioned. For many years the functionalist school has been dominating. However, Juengst has recently argued that it is necessary to supplement the functionalist approach with such an ontological view in order to reach an acceptable view on prevention and genetic enhancement (Juengst, 1997). According to the ontological approach, the causes of disease are detectable ontological entities such as genes, germs or environmental substances, and the pathological processes are 'real' in the same sense as properly functioning physiological systems. That the attack by such ontological entities through a pathological process is part of the objective dimension of the concept of disease means that it is a necessary condition for disease. If someone manifests disease, there must exist such ontological entities and such a pathological process. However, as with biological abnormality, this does not mean that the ontological entities and the pathological process can actually be detected. Even in this case, further research might be necessary.

It is obvious that on the basis of these two accounts–individually or together–the line between disease and nondisease is less fuzzy than that between serious and nonserious disease (and between certain nondisease traits and others). This is enough for policy purposes in reprogenetic medicine. Within the boundaries of disease understood in this way, the reproductive autonomy of prospective parents should be respected, i.e. they should be encouraged to make their own free choices. From the perspective of cognitive semantics, cases in line with the two accounts can be described as prototypical cases of disease. To be sure, there may also be controversial, nonprototypical cases. My argument does not imply that the line is not fuzzy at all, only that it is less fuzzy than that between serious and nonserious disease. A case often put forward is very short stature, which may have different causes. It may depend on a decreased level of human growth hormone or on the fact that the very short person has very short parents. Should very short stature be regarded as a disease? If the cause is a decreased level of human growth hormone, it is a disease. If the cause is very short parents, it is not a disease. This case shows that the line between disease and nondisease is fuzzy. But it does not show that there are serious problems of classification in general. It is only an exception, a nonprototypical case. Neither does it imply serious problems with regard to reprogenetic policy. Whether prospective parents should have reproductive autonomy in cases like these has to be judged on a case-by-case basis, as suggested by the approach of imaginative casuistry.

Further Remarks

Reproductive autonomy is not just a matter of the government or professionals of reprogenetic medicine not interfering with the procreative choices of prospective parents. It has also a positive side. As a matter of social justice, the government should, at least to some extent, have the goal to provide, within the limits presented above, certain genetic facilities for those who want to make use of them. In practice, however, certain permitted medical interventions can be denied with reference to fair distribution of limited resources. Scarcity of resources is a reality in all public health care systems. In such systems, it is wise to first offer genetic interventions on severe diseases and then, if financially possible, turn to less severe ones. In private health care, on the other hand, no medical interventions within the restrictions can be denied, if the prospective parents are prepared to pay for them.

Reproductive autonomy is the autonomy of prospective parents, i.e. for the prospective father and the prospective mother together. How should a conflict between these two parties be handled? If there is a conflict that cannot be resolved, for example regarding whether to carry out selective abortion, the woman should ultimately decide. The reason is that the fetus is a part of her body, not of the father's. However, the interests of the prospective father should always be taken seriously.

Let me also stress that a high appreciation of reproductive autonomy is not inconsistent with 'responsible reproduction' and a sensitivity for the needs of community. There are many things that can be done to ensure a responsible use of gene technology. The following examples may be noted.

It is extremely important that prospective parents be encouraged to moral reflection before they make their reproductive choices. This requires comprehensive information for prospective parents not only about risks but also about the uncertainty of predictions regarding the severity and time for onset of genetic diseases. They should also be informed of the fact that there are parents who think that having disabled children works out fine. Perhaps they should even be given opportunities to meet such parents. Prospective parents should also be informed that there are many disabled persons who think that they are living a life with a high quality in spite of their disability. In this way prejudices concerning the raising of disabled children might be corrected.

Moreover, if one stresses the principle of autonomy so as to allow certain kinds of parental eugenics, it is important that one at the same time tries to counteract discrimination of persons living with the disabilities that are selected against by certain prospective parents. It is also important to change the social environment to improve the situation of disabled persons. A disability can be defined as "a mismatch between the individual's abilities and the demands of a range of tasks", which means that there are fundamentally two ways of preventing disabilities, namely changing the individual or changing the social environment that defines the tasks (Buchanan, 1996, p. 39). In this chapter, I have discussed the former option, and stressed that phenotypic prevention is the only acceptable policy of reprogenetic medicine, never genotypic prevention. However, it is very important to realize that this policy is not inconsistent with one of improving the social environment for the disabled. Both are needed in society.

Finally, within the limits outlined above, threats to reproductive autonomy must be fought. In national health care systems with limited resources, politicians and health care officials might begin to apply pressure on prospective parents to prevent the birth of children with certain serious, and therefore costly, diseases and disabilities. The ambition of reprogenetic medicine should be to fight such pressures. Should the resources be too limited, it can reasonably be argued that they should be strengthened by increased taxes or more efficient planning. In private health care systems, insurance companies might also put pressure on prospective parents not to give birth to disabled children due to high costs. This should also be opposed by the geneticists. Moreover, in a society with a market economy, prospective parents might come under great pressure from biotechnological companies to use available reproductive technologies as much as possible. Beyond a certain limit pressures like these must also be counteracted. The professionals of reprogenetic medicine have a responsibility to ensure that the reproductive autonomy of prospective parents is not violated.

Two Objections to the Proposed View

A possible objection to this policy proposal is that if the genetics community has a professional policy of respecting reproductive autonomy within the boundaries of disease, this might lead to a kind of parental eugenics. Some prospective parents would use methods of genotypic prevention, for example selective abortion, in order to avoid giving birth to children with certain diseases. This would imply a devaluation of persons living with these diseases. By accepting such parental eugenics the government in fact says that persons living with such diseases have a lower value than other people. This would be a violation of the principle of equal human value.

This objection is not tenable, although it makes a valid point at the psychological level. First, we have to make a distinction between a disease on the one hand, and persons living with the disease on the other. It is completely possible to abort a fetus to prevent it from being born with a certain disease, and at the same time deeply care for persons living with this disease. In other words, it is possible to devalue a disease without devaluing persons living with it. For example, we devalue and try to prevent high blood pressure, without at the same time devaluing persons with high blood pressure. Similarly, we can devalue cystic fibrosis or Down's syndrome without devaluing persons with these conditions.

Second, we have to make a distinction between embryos and fetuses, on the one hand, and persons living with a disease, on the other. It is possible to argue that while persons living with a disease have full moral standing, embryos and fetuses do not. Persons living with a disease therefore have a right not to be killed while under certain circumstances it might be morally acceptable to abort embryos or fetuses. This is a view shared by all but strict opponents to all kinds of abortion. However, this is not the place for arguing extensively in favor of it.

These remarks show that it is possible to carry out a selective abortion on a fetus with genes for a certain disease without expressing a devaluation of persons living with this disease. And it is possible for the government to allow selective abortion without expressing such a devaluation. My conclusion is therefore that this objection to my proposal is not logically compelling. However, the argument carries some conviction at a psychological level. It points out a possible psychological effect of selective abortion. Persons with the disease as well as persons without the disease might interpret selective abortion as a devaluation of those living with the disease. It is important that health care have the policy to clearly take a stand against such interpretations and show in practice that it is possible both to allow selective abortion and care for those with the diseases.

Another possible objection to the proposal is that it is one thing to argue that genetic interventions on nondisease traits of the embryo or fetus lie outside the domain of health care, but quite another to argue that they should not be allowed at all in society. Why not allow nonmedical institutions to make such interventions? In our present society, we already allow a vast number of nongenetic interventions concerning such traits, e.g. physical education, training of cognitive abilities, and cosmetic surgery, and one might wonder why genetic interventions could not be used to achieve similar goals.

However, giving some people too great advantages already from the start of their lives by nondisease genetic interventions would probably deviate too greatly from values appreciated by the majority. Moreover, such interventions would probably for a foreseeable future be an opportunity only for the rich. We would see a 'new genetic upper class' born with far better potentials than other people. This might imply discrimination and stigmatization of the 'unselected' or 'unenhanced'. Such a situation is not compatible with the principle of justice and would probably also create too much tension in society. In addition, if certain prospective parents choose to enhance the aggressiveness of their children in order to help them in an increasingly competitive society, this would be even more undesirable.

The Long Term Goals of Reprogenetic Medicine

What standpoint regarding the long term goals of reprogenetic medicine should the professional policy of the genetics community reasonably take? Let me briefly summarize a few suggestions for consideration.

Diagnosis and prognosis. Diagnosis is, of course, an important goal. Genetic testing should, generally, only be carried out for diseases, not for nondisease traits. However, if prospective parents request information about the sex of the embryo or fetus, this could be approved if there is no indication that this information will be used for selective purposes but is only requested out of pure curiosity.

Treatment and cure. A second goal of reprogenetic medicine should be to offer treatment to individuals for genetic diseases using, for instance, genetically tailor-made drugs and gene therapy. The main justification for this is the principle of beneficence. Research and clinical use of somatic cell gene therapy on fetuses is acceptable given adequate safety measures. Research on germline gene therapy should not be permitted at present. It can be accepted in principle, but only on certain very strict conditions. These can be summarized as follows.

(a) Germline gene therapy is the only or most efficient means of achieving successful treatment of the embryo.

(b) The safety of somatic cell gene therapy is very well established.

(c) The risks of carrying out germline gene therapy research are predicted to be minor.

(d) Germline gene therapy is carried out at the request of the prospective parents, not of the government, the professionals of reprogenetic medicine, or other agents.

These conditions are not at all fulfilled at present, and will likely not be in the foreseeable future. This means that germline gene therapy research should not be started. However, if the conditions for research are fulfilled sometime in the future and the technology becomes established and sufficiently safe, clinical use of germline gene therapy could be accepted.

Prevention. A third goal of reprogenetic medicine should be prevention, but only phenotypic prevention, not genotypic. Whatever measures are used, they must "benefit directly those involved", as the *Declaration of Inuyama* states (CIOMS, 1995). This means that the use of germline gene therapy in public health programs for preventing diseases in future generations should not be accepted, even if the conditions for carrying out germline gene therapy research are fulfilled and a safe technology is developed. Such governmental and professional steering of reproduction would imply improper pressure on prospective parents, and, therefore, a violation of reproductive autonomy. No such public health eugenics should be accepted, not even a voluntary one that gives prospective parents the right to say 'no'. Reproductive autonomy should be interpreted more radically than that. Public health work should only include phenotypic prevention.

However, genotypic prevention as a measure autonomously chosen by prospective parents should be accepted, at least within certain limits. Nondisease-related parental eugenics should be discouraged. Moreover, if germline gene therapy becomes developed sometime in the future, it could be used by prospective parents as a method for genotypic prevention. If prospective parents at that future point of time want a genetic disease to be eliminated for good from their family line, it would be acceptable to do so by means of this technology, although the primary aim should always be to help the embryo at hand. Until germline gene therapy is established, however, selective abortion and, when it becomes generally available, preimplantation selection, are the only methods to be used for genotypic prevention.

Enhancement. When a sufficient degree of safety is reached in research on somatic cell gene therapy, research on disease-centered somatic cell genetic enhancement on the fetus, for instance for improving the immune system, should be accepted. Clinical use of this disease-centered genetic enhancement should be accepted only when its safety is very well established. Provided that germline gene therapy is developed, disease-centered germline genetic enhancement could also be accepted, but only if the primary goal is to help the embryo itself, not future generations, and only if such an intervention is the only way to do so. Moreover, somatic and germline enhancement should be conducted only at the request of prospective parents.

Reproductive autonomy. This should be not just a principle for the implementation of the goals of reprogenetic medicine but a goal in itself, at least in the sense that increased reproductive autonomy is valuable. It should be noted, however, that the view that increased autonomy should be a goal holds for reprogenetic medicine but not necessarily for medicine in general. For reasons given above, reproduction is special. The negative aspect of this reproductive autonomy is that prospective parents should be left to decide on their own. But there should be certain restrictions, namely that the reproductive autonomy should hold only within the boundaries of disease.

Fair access. The positive aspect of reproductive autonomy is that the government should provide at least some means to ensure that this autonomy does not merely become a privilege for the rich. This is actually the same as to say that there should be fair access to genetic facilities which is something that the principle of justice demands.

Implementation of the Goals: Forms of Responsibility

How should these goals be integrated with scientific and clinical practice? What should be the means of implementing the goals? These are questions about the forms of responsibility (see pp. 89-90, 216). The professional policy of the genetics community should include not only statements about the goals of reprogenetic medicine but also those about the forms of responsibility. In what ways should the content of responsibility be realized?

First, it is necessary to distinguish between research and application.[9] What are proper forms of responsibility with respect to the former may not be proper with respect to the latter.

With regard to research, let us focus on germline gene therapy research as an example. I have suggested that it is not acceptable to start this research at present. How should this policy be implemented? There are several possibilities, each with its advantages and disadvantages.[10] One alternative is to develop specific professional guidelines for the researchers. These guidelines could hold nationally or globally. An advantage is that this alternative takes seriously the moral responsibility of the genetics community, and if the geneticists keep to the guidelines this would give the geneticists great credibility and high moral status in society. A disadvantage, however, is that it is improbable that all researchers will keep to the guidelines. Out of pure curiosity or for commercial reasons, some scientists may nevertheless want to carry out this kind of research. A more efficient alternative could be governmental recommendations or those by the EU or the UN. This may put more pressure on geneticists not to start doing the research. On the other hand, mere recommendations may not be sufficient to stop them. This brings us to the alternative of legal regulation or a legal moratorium. We already have laws against germline gene therapy in many countries. This form of responsibility might be effective in a specific country, at least for a while. But there is always a possibility that the research will be carried out in another country. If we truly want

[9] With regard to application, it may also be important in a more elaborate presentation to distinguish between the introduction of a specific genetic technology into health care and the established use of the technology in health care (cf. Munthe, 1999, p. 246). For my more restricted purposes, however, the distinction is not necessary.

[10] A few existing policy documents with regard to germline gene therapy are presented on pp. 183-184.

to prevent germline gene therapy research globally, this may not be efficient, either. Finally, we have the alternative of steering research by *not* providing any public funding. The government simply refrains from financing this kind of research. The risk is, of course, that this makes germline gene therapy research even more interesting for private funding.

It is not easy to choose between these alternatives, and, of course, much dialogue is needed. Since this is a matter of efficiency, it may be wise to try different alternatives. Moreover, a combination of methods may be necessary. If possible, legal regulation should be avoided. But with regard to germline gene therapy research, it may be necessary. Global guidelines and recommendations may also be needed. Note, however, that I am not against germline gene therapy in principle. This means that I keep open the possibility that sometime in the future it might become reasonable to change the laws, recommendations, and guidelines, and permit germline gene therapy research.

If we turn to the professional policy on application, let us take the example of providing genetic testing and selection procedures only within the boundaries of disease. I have suggested that the geneticists should discourage prospective parents from seeking genetic information about nondisease traits and carrying out selection on the basis of this information. How can this be implemented in practice? Specific professional guidelines stating the importance of discouraging prospective parents may be efficient to some extent. But not only may there be problems of convincing all professionals of reprogenetic medicine to accept the guidelines, it may be even more difficult to convince all prospective parents that this policy is justified. Governmental recommendations or even legal regulation could be alternatives. However, I find both alternatives problematic as they imply too much governmental involvement in procreation.

In conclusion, it seems that the best alternative is specific professional guidelines. The professionals of reprogenetic medicine should in the first step discourage prospective parents from seeking genetic information about nondisease traits and carrying out selection on the basis of this information. If some prospective parents want to proceed in spite of this, the professionals may, in the second step, refuse. This might lead the prospective parents to turn to other professionals who do not conform to the professional guidelines, but it might be necessary to accept this. Perhaps the policy of restricting reproductive autonomy should not be overly ambitious.

Conclusion

These suggestions regarding the long term goals of reprogenetic medicine and their implementation are only intended as a contribution to an interdisciplinary dialogue. The genetics community has to formulate its own professional policy. Doing so, however, is one of the most important ways of taking moral responsibility for the consequences of human genetics research.

BIBLIOGRAPHY

Agius E., 1990. "Germ-line cells: Our responsibilities to Future Generations." In: Busuttil E., Agius E., Inglott P.S., Macelli T. (eds.). *Our Responsibilities Towards Future Generations*. Foundation for International Studies: Villeta, Malta.

Agius E., Busuttil S. (eds.), 1998. *Germ-Line Intervention and Our Responsibilities to Future Generations*. Philosophy and Medicine Vol. 55. Kluwer Academic Publishers: Dordrecht.

Andrews L.B., 2000. *The Clone Age: Adventures in the New World of Reproductive Technology*. Henry Holt and Company: New York.

Annas G.J., Elias, S. (eds.), 1992. *Gene Mapping: Using Law and Ethics as Guides*. Oxford University Press: Oxford.

Annas G.J., Grodin, M.A. (eds.), 1992. *The Nazi Doctors and the Nuremberg Code: Human Rights in Human Experimentation*. Oxford University Press: Oxford.

Arras J.D., 1991. "Getting Down to Cases: The Revival of Casuistry in Bioethics." *The Journal of Medicine and Philosophy* 1991, 16, pp. 42-44.

Asch A., 1989. "Reproductive Technology and Disability." In: Cohen, S., Taub, N. (eds.), 1989. *Reproductive law for the 1990's*. Humana Press: Clifton.

Asch A., 1996. "Genetics and Employment: More Disability Discrimination." In: Murray, Rothstein, Murray (eds.), 1996.

Audi R. (general ed.), 1995. *The Cambridge Dictionary of Philosophy*. Cambridge University Press: Cambridge.

Baier A., 1984. "For the Sake of Future Generations". In: Regan T. (ed.), 1984. *Earthbound: New Introductory Essays in Environmental Ethics*. Random House: New York.

Barbour I.G., 1992. *Ethics in an Age of Technology*. SCM Press: London.

Barkow J.H., Cosmides L., Tooby J., 1992. *The Adapted Mind: Evolutionary Psychology and the Generation of Culture*. Oxford University Press: New York.

Bateson P., Martin P., 2000. *Design for a Life: How Biology and Psychology Shape Human Behavior*. Simon & Schuster: New York.

Bayles M. (ed.), 1976. *Ethics and Population Policy*. Shenkman: Cambridge.

Bayertz K., 1994. *GenEthics: Technological Intervention in Human Reproduction as a Philosophical Problem*. Cambridge University Press: Cambridge.

Beauchamp T.L., Childress J.F., 1989 (3rd ed.). *Principles of Biomedical Ethics*. Oxford University Press: New York.

Beauchamp T.L., Childress J.F., 1994 (4th ed.). *Principles of Biomedical Ethics*. Oxford University Press: New York.

Beck U., 1992. *Risk society: Towards a new modernity*. Sage: London.

Beckwith J., 1993. "A historical view of social responsibility in genetics." *BioScience* 43, 5, pp. 327-333.

Beckwith J., 1997. "The Responsibilities of Scientists in the Genetics and Race Controversies." In: Smith, Sapp (eds.), 1997.

Benton T., 1996. "Animal rights and social practices." In: Thomasma, Kushner (eds.), 1996.

Bereano P., 1997. "Technocracy and Democracy." In: Smith, Sapp (eds.), 1997.

Biesecker B.B., Marteau T.M., 1999. "The future of genetic counselling: an international perspective." *Nature Genetics* Vol. 22, pp. 133-137.

Biotechnology and the European Public Concerted Action Group, 1997. "Europe ambivalent on biotechnology." *Nature* 387, 26 June, pp. 845-847.

Bishofberger E., 1990. "The Ideas of Integrity and Autonomy: Discussion." In: Allebeck P., Jansson B. (eds.). *Ethics in Medicine*. Raven Press: New York.

Bishop J.E., Waldholz M., 1990. *Genome: The Story of the Most Astonishing Scientific Adventure of Our*

Time–The Attempt to Map all the Genes in the Human Body. Simon & Schuster: New York.
Blumstein J.F., 1996. "Legalizing payment for transplantable cadaveric organs." In: Thomasma, Kushner (eds.), 1996.
Boorse C., 1977. "Health as a theoretical concept." *Philosophy of Science* 44, pp. 542-573.
Botstein D., 1997. "Of Genes and Genomes." In Smith, Sapp (eds.), 1997.
Brenner M., 1999. "Gene Therapy: At Present and in the Future." In: Nordgren (ed.), 1999.
British Medical Association, 1998. *Human Genetics: Choice and Responsibility*. Oxford University Press: Oxford.
Broberg G., Roll-Hansen N., (eds.), 1996. *Eugenics and the Welfare State: Sterilization Policy in Denmark, Sweden, Norway, and Finland*. Michigan State University Press: East Lansing.
Brody B.A., 1988. *Life and Death Decision Making*. Oxford University Press: Oxford.
Brody B.A., 1998. *The Ethics of Biomedical Research: An International Perspective*. Oxford University Press: Oxford.
Buchanan A., 1996. "Choosing Who Will Be Disabled: Genetic Intervention and the Morality of Inclusion." In: Frankel Paul, Miller, Paul (eds.), 1996.
Bulger R.E., Heitman E., Reiser S.J. (eds.), 1993. *The Ethical Dimensions of the Biological Sciences*. Cambridge University Press: Cambridge.
Callahan D., 1976. "Ethical responsibility in science in the face of uncertain consequences." In: Lappé, M., Morison, R.S. (eds.). *Ethical and scientific issues posed by human uses of molecular genetics*. The New York Academy of Sciences: New York.
Caplan A., 1993. "Beastly conduct: Ethical Issues in animals experimentation." In: Bulger, Heitman, Reiser (eds.), 1993.
Capron A.M., 1990. "Which Ills to Beat: Reevaluating the 'Threat' of Modern Genetics." *Emory Law Journal* 39, pp. 665-696.
Caulfield T., Robertson G., 1996. "Eugenic Policies in Alberta." *Alberta Law Review* XXXV, 1, pp. 59-79.
Cavalli-Sforza L.L., 1997. "Race Differences: Genetic Evidence." In: Smith, Sapp (eds.), 1997.
Cavazzana-Calvo M. et al., 2000. "Gene therapy of human severe combined immunodeficiency (SCID)-X1 disease." *The Journal of Gene Medicine*, Supplement to Vol. 2 No. 5, p. 16.
Chadwick R. (ed.), 1992. *Ethics, Reproduction and Genetic Control*. Routledge: London.
Chadwick R.F. (ed.), 1999. *The Ethics of Genetic Screening*. Kluwer Academic Publishers: Dordrecht.
Chan A.W.S., Chong K.Y., Martinovich C., Simerly C., Schatten G., 2001. "Transgenic Monkeys Produced by Retroviral Gene Transfer into Mature Oocytes." *Science* 291, pp. 309-312.
Childress J.F., 1995. "Metaphor and analogy." In: Reich W.T. (ed.), 1995. *Encyclopedia of Bioethics*, pp. 1765-1773.
CIOMS (Council for International Organisations of Medical Sciences), 1995. *The Declaration of Inuyama: Human Genome Mapping, Genetic Screening and Gene Therapy*. CIOMS: Geneva.
Clarke A. (ed.),1994. *Genetic Counselling: Practice and Principles*. Routledge: London.
Clayton E.W. et al., 1995. "Informed Consent for Genetic Research on Stored Tissue Samples." *JAMA* 274, 22, pp. 1786-1792.
Clouser K.D., Culver C., Gert B., 1981. "Malady: A new treatment of disease." *Hastings Center Report* 11, pp. 29-37.
Cohen C., 1994. "The Case for the Use of Animals in Biomedical Research." In: Erwin, Gendin, Kleiman (eds.), 1994.
Commission of the European Communities, 1988. *Proposal for a Council Decision Adopting a Specific Research Programme in the Field of Health; Predictive Medicine: Human Genome Analysis (1989-1991)*. COM (88) 424 final-SYN, Brussels, 20 July 1988.
Commission of the European Communities, 1989. *Modified Proposal for a Council Decision Adopting a Specific Research and Technological Development Programme in the Field of Health: Human Genome Analysis (1990-1991)*. 13 November 1989.
Cook-Deegan R., 1995 (1994). *The Gene Wars: Science, Politics, and the Human Genome*. Norton: New York.
Council Directive on the approximation of laws, regulations and administrative provisions of the Member

States regarding the protection of animals used for experimental and other scientific purposes, 1986. 86/609/EEC.
Council of Europe, Parliamentary Assembly, 1982. Report on genetic engineering presented by the Legal Affairs Committee. Document 4832 of the 33rd Ordinary Session. Strasbourg.
Council of Europe, 1996. *Convention on Human Rights and Biomedicine*. Directorate of Legal Affairs, Strasbourg.
Daniels N., 1979. "Wide reflective equilibrium and theory acceptance in ethics." *The Journal of Philosophy* 76, pp. 256-282.
De Wachter M.A.M., 1993. "Ethical Aspects of Human Germ-Line Gene Therapy." *Bioethics* 7, pp. 166-177.
Dierickx K., 1999. "The Impact of Human Gene Therapy on Genetic Screening and Testing: Ethical and Societal Considerations." In: Nordgren (ed.), 1999.
Directive of the European Parliament and the Council of 6 July 1998 on the legal protection of biotechnological inventions. Official Journal L 213, 30/07/1998, pp. 0013-0021.
Duster T., 1990. *Backdoor to Eugenics*. Routledge: New York.
Dyson A., Harris J. (eds.), 1994. *Ethics and Biotechnology*. Routledge: London.
Eisenberg R.S., 1996. "Patent Rights in the Human Genome Project." In: Annas, Elias (eds.), 1996.
Elliott D., Stern J.E. (eds.), 1997. *Research Ethics: A Reader*. University Press of New England: Hanover.
ELSI, 1993. *Genetic Information and Health Insurance: Report of the Task Force on Genetic Information and Insurance*. National Institutes of Health: Bethesda.
ELSI, 1996. "ELSI Working Group responds to The Bell Curve." *Human Genome News*, January–March, 16.
Emanuel E.J., 1991. *The Ends of Human Life: Medical Ethics in a Liberal Polity*. Harvard University Press: Cambridge.
Emanuel E., Emanuel L., 1995. "Four Models of the Physician-Patient Relationship." In Arras J., Steinbock B. (eds.). *Ethical Issues in Modern Medicine*. Mayfield Publishing Company: Mountain View.
Emilien G., Ponchon M., Caldas C., Isacson O., Maloteaux J.-M., 2000. "Impact of genomics on drug discovery and clinical medicine." QJM 93, pp. 391-423.
Engelhardt H.T. Jr, 1986. *The Foundations of Bioethics*. Oxford University Press: Oxford.
Erin C.A., 1994. "Who Owns MO? Using historical entitlement theory to decide the ownership of human derived cell lines." In: Dyson, Harris (eds.), 1994.
Erwin E., Gendin S., Kleiman L. (eds.), 1994. *Ethical Issues in Scientific Research*. Garland Publ.: New York.
European Community, 1989. *Common Position Adopted by the Council on the 15 December 1989: A Specific Research and Technological Development Programme in the Field of Health: Human Genome Analysis (1990-1991)*. Brussels 15 December 1989,10619/89.
Feinberg J., 1970. *Doing and Deserving*. Princeton University Press: Princeton.
Flynn J.R., 1980. *Race, IQ, and Jensen*. Routledge: London.
Frankel M.S. (ed.), 1997. *Exploring Public Policy Issues in Genetics*. AAAS: Washington D.C.
Frankel M.S., Teich A.H. (eds.), 1993. *Ethical and Legal Issues in Pedigree Research*. AAAS: Washington D.C.
Frankel Paul E., Miller Jr., F.D., Paul J. (eds.), 1996. *Scientific Innovation, Philosophy, and Public Policy*. Cambridge University Press: Cambridge.
Frankena W., 1973 (1963). *Ethics*. Prentice-Hall: Englewood Cliffs.
French, P.A., 1992. *Responsibility Matters*. University Press of Kansas: Lawrence.
Gifford F., 1994. "The Conflict between Randomized Clinical Trials and the Therapeutic Obligation." In: Erwin, Gendin, Kleiman (eds.), 1994.
Gillon R., 1986. *Philosophical Medical Ethics*. Wiley: Chichester.
Glover J., 1984. *What Sort of People Should There Be?* Penguin: Harmondswoth.
Glover J., 1970. *Responsibility*. Humanties Press: London.
Gould S.J., 1996. *The Mismeasure of Man*. Penguin: London.
Graumann S., 1999. "Some Conceptual Questions about Somatic Gene Therapy and their Relevance for

an Ethical Evaluation." In: Nordgren (ed.), 1999.
Gross P.R., Levitt N., 1994. *Higher Superstition: The Academic Left and Its Quarrels with Science*. The Johns Hopkins University Press: Baltimore.
Grove J.W., 1989. *In Defence of Science: Science, Technology, and Politics in Modern Society*. University of Toronto Press: Toronto.
Gustafsson B., Rydén L., Tibell G., Wallensteen P., 1984. "The Uppsala Code of Ethics for Scientists." *Journal of Peace Research* 21, 4, pp. 311-316.
Haker H., 1993. "Human Genome Analysis and Eugenics." In: Haker H., Hearn R., Steigleder K., (eds.), 1993. *Ethics of Human Genome Analysis: European Perspectives*. Francke: Tübingen.
Hallen M. (ed.), 1998. *Human Genome Analysis supported under BIOMED 1*. IOS Press: Amsterdam.
Hallen M., Klepsch A. (eds.), 1995. *Human Genome Analysis Programme*. IOS Press: Amsterdam.
Hamer D., Copeland P., 2000 (1998). *Living with Our Genes: Why Genes Matter More than You Think*. Pan Books: London.
Hanson M.J., Callahan D., 1999. *The Goals of Medicine: The Forgotten Issues in Health Care Reform*. Georgetown University Press: Washington D.C.
Harris, J., 1998. *Clones, Genes, and Immortality: Ethics and the Genetic Revolution*. Oxford University Press: Oxford.
Hart H.L.A., 1968. *Punishment and Responsibility*. Oxford University Press: Oxford.
Heeger R., 1997. "Respect for Animal Integrity?" In: Nordgren (ed.), 1997a.
Herrnstein R.J., Murray C. (1994). *The Bell Curve. Intelligence and Class Structure in American Life*. The Free Press: New York.
Hoedemaekers R., ten Have H., 1998. "Geneticization: The Cyprus Paradigm." *Journal of Medicine and Philosophy* 23, 3, pp. 274-287.
Holloway M., 1999. "Flynn's Effect." *Scientific American* 280, 1, pp. 20-21.
Holtug N., 1997. "Altering Humans: The Case For and Against Human Gene Therapy." *Cambridge Quarterly of Healthcare Ethics* 6, pp. 157.174.
Hubbard R., Wald E., 1993. *Exploding the Gene Myth. How Genetic Information Is Produced and Manipulated by Scientists, Physicians, Employers, Insurance Companies, Educators, and Law Enforcers*. Beacon Press, Boston.
International Human Genome Consortium, 2001. "Initial sequencing and analysis of the human genome." *Nature* Vol. 409 (15 February), pp. 860-921.
Jackson F., 1997. "Assessing the Human Genome Project: An African American and Bioanthropological Critique." In: Smith, Sapp (eds.), 1997.
Jecker N., 1997. "Introduction to the Methods of Bioethics." In: Jecker, Jonsen, Pearlman (eds,), 1997.
Jecker N., Jonsen A., Pearlman R. (eds.), 1997. *Bioethics: An Introduction to the History, Methods, and Practice*. Jones and Bartlett Publishers: Boston.
Johnson M., 1987. *The Body in the Mind*. Universtiy of Chicago Press: Chicago.
Johnson M., 1993. *Moral Imagination: Implications of Cognitive Science for Ethics*. University of Chicago Press: Chicago.
Johnson M.L., 1996. "How Moral Psychology Changes Moral Theory." In: May, Friedman, Clark (eds.), 1996.
Jonas H., 1984. *The Imperative of Responsibility: In Search of an Ethics for the Technological Age*. Chicago University Press: Chicago.
Jonsen A.R., 1968. *Responsibility in Modern Religious Ethics*. Corpus Books, Washington D.C.
Jonsen A.R., Toulmin S., 1988. *The Abuse of Casuistry: A History of Moral Reasoning*. University of California Press: Berkeley.
Jonsen A.R., Siegler M., Winslade W.J., 1998 (1982). *Clinical Ethics: A Practical Approach to Ethical Decisions in Clinical Medicine*. McGraw-Hill: New York.
Juengst E.T., 1994. "Respecting Human subjects in Genome Research: A Preliminary Policy Agenda." In: Vanderpool, H.Y. (ed.). *The Ethics of Research Involving Human subjects: Facing the 21st Century*. University Pub. Group: Frederick.
Juengst E.T., 1995a. "The Ethics of Prediction: Genetic Risk and the Physician–Patient Relationship." *Genome Science & Technology* 1, pp. 21-36.

Juengst E.T., 1995b. "'Prevention' and the Goals of Genetic Medicine." *Human Gene Therapy* 6, pp. 1595-1605.
Juengst E.T., 1996. "Self-Critical Federal Science? The Ethics Experiment within the U.S. Human Genome Project." In: Frankel Paul, Miller, Paul (eds.), 1996.
Juengst, E.T., 1997. "Can Enhancement Be Distinguished from Prevention in Genetic Medicine?" *The Journal of Medicine and Philosophy* 22, pp. 125-142.
Juengst E.T., 1998a. "Groups as Gatekeepers to Genomic Research: Conceptually Confusing, Morally Hazardous, and Practically Useless." *Kennedy Institute of Ethics Journal* 8, pp. 183-200.
Juengst E.T., 1998b. "Should we treat the human germ-line as a global human resource?" In: Agius, Busuttil (eds.), 1998.
Juengst E.T., 1999. "Anticipating Enhancement: A Conceptual and Ethical Challenge for Gene Therapy Regulation." In: Nordgren (ed.), 1999.
Kaiser M.,1993. "Some Thoughts on the Responsibility of Scienctists in Relation to the Growth of Fish-Farming in Norway." In: Welin (ed.), 1993.
Karp L.E., 1983. "The terrible question." *American Journal of Medical Genetics* 23, pp. 359-362.
Kass L., 1972. "New Beginnings in Life". In: Hamilton M. (ed.). *The New Genetics*. Eerdmans: Grand Rapids, MI.
Kaye H.L., 1997. *The Social Meaning of Modern Biology: From Social Darwinism to Sociobiology*. Transaction Publishers: New Brunswick.
Keller E.F., 1995. *Refiguring Life: Metaphors of Twentieth-Century Biology*. Columbia University Press: New York.
Kevles D.J., 1992. "Out of Eugenics: The Historical Politics of the Human Genome." In: Kevles, Hood (eds.), 1992.
Kevles D.J., 1994. "Eugenics and the Human Genome Project: Is the Past Prologue?" In: Murphy, Lappé (eds.), 1994.
Kevles D.J. 1995 (4th ed., with a new preface) (1985). *In the Name of Eugenics: Genetics and the Uses of Human Heredity*. Harvard University Press: Cambridge.
Kevles D.J., Hood, L. (eds.), 1992. *The Code of Codes: Scientific and Social Issues in the Human Genome Project*. Harvard University Press: Cambridge, MA.
Khoury M.J. et al, 1998. "Translating Advances in Human Genetics into Disease Prevention and Health Promotion." *Community Genetics* 1, pp. 93-98.
Kimura R., 1991. "Jurisprudence in Genetics." In: Srám R.J., Bulyzhenkov V., Prilipko L., Christen Y. (eds.). *Ethical Issues of Molecular Genetics in Psychiatry*. Springer-Verlag: Berlin.
Kingsbury D.T., 1997. "A General View of the Human Genome Project." In: Frankel (ed.), 1997.
Kitcher P., 1996. *The Lives to Come. The Genetic Revolution and Human Possibilities*. Allen Lane, The Penguin Press: New York.
Kumar D., 1985. "Should One Be Free to Choose the Sex of One's Child?" In: Chadwick (ed.), 1992.
Lakoff G., 1990 (1987). *Women, Fire, and Dangerous Things: What Categories Reveal about theMind*. University of Chicago Press: Chicago.
Lakoff G., Johnson M., 1980. *Metaphors We Live By*. University of Chicago Press: Chicago.
Lakoff G., Johnson M., 1999. *Philosophy in the Flesh: The Embodied Mind and Its Challenge to Western Thought*. Basic Books: New York.
Lappé M., 1995. "Eugenics." In: Reich W.T. (ed.). *Encyclopedia of Bioethics*. Macmillan: New York, pp. 770-777.
Lebacqz K., 1999. "The Ambiguities of 'Therapy' and their Implications for the Place of Women and the Disabled in a Gene-Focused Society." In: Nordgren (ed.), 1999.
Levinas E., 1985 (French original: 1982). *Ethics and Infinity: Conversations with Philippe Nemo*. Duquesne University Press: Duquesne.
Levine R.J., 1996. "Research involving children." In: Thomasma, Kushner (eds.), 1996.
Lewontin R.C., 1992. *The doctrine of DNA: Biology as Ideology*. Penguin: New York.
Lippman A., Wilfond B.S., 1992. "Twice-told tales: stories about genetic disorders." *American Journal of Human Genetics* 51, pp. 936-937.
Lucas J.R., 1995. *Responsibility*. Clarendon Press: Oxford.

MacIntyre A., 1985 (1981). *After Virtue: A Study in Moral Theory*. Duckworth: London.
Mao X., Wertz D.C., 1997. "China's genetic services providers' attitudes towards several ethical issues: a cross-cultural study." *Clinical Genetics* 52, pp. 100-109.
Marteau T., Richards M. (eds.), 1996. *The troubled helix. Social and psychological implications of the new human genetics*. Cambridge University Press: Cambridge.
Maturana H., Varela F, 1992 (rev. ed.). *The Tree of Knowledge: The Biological Roots of Human Understanding*. Shambala: Boston, London.
May L., 1992. *Sharing Responsibility*. The University of Chicago Press: Chicago.
May L., Friedman M., Clark A. (eds.), 1996. *Mind and Morals: Essays on Cognitive Science and Ethics*. The MIT Press: Cambridge
McKusick V.A., 1989. "The Human Genome Organisation: History, Purposes and Membership." *Genomics* 5, pp. 385-387.
Mepham T.B. et al, 1998. "The Use of Transgenic Animals in the European Union: The Report and Recommendations of ECVAM Workshop 28." *ATLA* 26, pp. 21-43.
Merges R.P., 1996. "Property Rights Theory and the Commons: The Case of Scientific Research." In: Frankel Paul, Miller, Paul (eds.), 1996.
Merton R., 1973. "The Normative Structure of Science" (1st ed. of the paper: 1942). In: Merton R. *The Sociology of Science*. University of Chicago Press: Chicago.
Midgley M., 1995. "Duties concerning islands." In: Elliot R. (ed.), *Environmental Ethics*. Oxford University Press: Oxford.
Mieth D., 1999. "The Ethics of Gene Therapy: The German Debate." In: Nordgren (ed.), 1999.
Moran G., 1996. *A Grammar of Responsibility*. The Crossroad Publishing Company: New York.
Moros D.A., 1996. "Taking duties seriously: Medical experimentation, animals rights, and moral incoherence." In: Thomasma, Kushner (eds.), 1996.
Munson R., Davis L.H., 1992. "Germ-Line Gene Therapy and the Medical Imperative." *Kennedy Institute of Ethics Journal* 2, 2, pp. 137-158.
Munthe C., 1996. *The Moral Roots of Prenatal Diagnosis: Ethical Aspects of the Early Introduction and Presentation of Prenatal Diagnosis in Sweden*. Studies in Research Ethics No. 7. Centre for Research Ethics: Göteborg.
Munthe, C., 1999. *Pure Selection: The Ethics of Preimplantation Genetic Diagnosis and Choosing Children without Abortion*. Acta Philosophica Gothoburgensia 9. Acta Universitatis Gothoburgensis: Göteborg.
Murashige K.H., 1997. "The Human Genome Project and the Patent System–Is There a Useful Nexus?" In: Frankel M.S. (ed.), 1997.
Murphy T.L., Lappé M.A. (eds.), 1994. *Justice and the Human Genome Project*. University of California Press: Berkeley.
Murray T.H., 1992. "The Human Genome Project and Genetic Testing: Ethical Implications." In: AAAS. *The Genome, Ethics and the Law*, pp. 49-78.
Murray T.H., 1996. *The Worth of a Child*. University of California Press: Berkeley.
Murray T.H., Rothstein M.A., Murray R.F. (eds.), 1996. *The Human Genome Project and the Future of Health Care*. Indiana University Press: Bloomington.
National Center for Human Genome Research, 1991. *The Human Genome Project: New Tools for Tommorow's Health Research*. National Institutes of Health: Bethesda, MD.
National Human Genome Research Institute, 1997. *Progress Report Fiscal Years 1995-1996: The Human Genome Project*. National Institutes of Health: Bethesda, MD.
Neisser U. (ed.), 1998. *The Rising Curve: Lond-Term Gains in IQ and Related Issues*. American Psychological Association: Washington D.C.
Nelkin D., 1994. "Promotional metaphors and their popular appeal." *Public Understand. Sci.* 3, pp. 25-31.
Nelkin D., Lindee, M.S., 1995. *The DNA Mystique. The Gene as a Cultural Icon*. WH Freeman and Company: New York.
Niebuhr H.R., 1963. *The Responsible Self: An Essay in Christian Moral Philosophy*. Harper & Row: New York.

Nordenfelt L., 1995 (2nd rev. ed.) (1987). *On the Nature of Health: An Action-Theoretic Approach.* Philosophy and Medicine series Vol. 26. Kluwer Academic Publishers: Dordrecht.
Nordgren A, 1994. *Evolutionary Thinking. An Analysis of Rationality, Morality and Religion from an Evolutionary Perspective.* A&W International: Stockholm.
Nordgren, A. (ed.), 1997a. *Science, Ethics, Sustainability. The Responsibility of Science in Attaining Sustainable Development.* Studies in Bioethics and Research Ethics 2. Acta Universitatis Upsaliensis: Uppsala.
Nordgren A., 1997b. "Science and Sustainability. Some Reflections on the Moral Responsibility of Scientists." In: Nordgren (ed.), 1997a.
Nordgren A., 1998a. "Ethics and Imagination: Implications of Cognitive Semantics for Medical Ethics." *Theoretical Medicine and Bioethics*, 19, 2, pp. 117-141.
Nordgren A., 1998b. "Reprogenetics Policy: Three Kinds of Models." *Community Genetics* 1, pp. 61-70.
Nordgren A. (ed.), 1999. *Gene Therapy and Ethics.* Studies in Bioethics and Research Ethics 4. Acta Universitatis Upsaliensis: Uppsala.
Nordgren A., Westrin C.-G. (eds.), 1998. *Altruism, Society, Health Care.* Studies in Bioethics and Research Ethics 3. Acta Universitatis Upsaliensis: Uppsala.
Nussbaum M., 1986. *The Fragility of Goodness.* Cambridge University Press: Cambridge.
Official Journal of the European Communities, 1990. No. L 196/8, 26/7/90. Council Decision of 29 June 1990. Adopting a specific research and technological development programme in the field of health, human genome analysis (1990-1991), (90/395/EEC).
Orlans F.B., 1993. *In the name of Science: Issues in Responsible Animal Experimentation.* Oxford University Press: Oxford.
Parfit D., 1976. "On Doing the Best for Our Children." In: Bayles (ed.), 1976.
Paul D.B., 1991. "Book Review: Enthusiastic Claims." *Science* 252, pp.142-143.
Paul D.B., 1994. "Is Human Genetics Disguised Eugenics?" In: Weir, Lawrence, Fales, (eds.), 1994.
Paul D., 1997. "From Eugenics to Medical Genetics." *The Journal of Policy History* 9, pp. 96-116.
Paul D.B., Spencer, H.G., 1995. "The hidden science of eugenics." *Nature* 374, pp. 302-304.
Pedersen N.L., Plomin R., Nesselroade J.R. "Individual differences in locus of control during the second half of the life span for identical and fraternal twins reared apart and reared together." *Journal of Gerontology* 44, pp. 100-105.
Pejovich S., 1996. "Property Rights and Technological Innovation." In: Frankel Paul, Miller, Paul (eds.), 1996.
Pellegrino E.D., 1990. "The relationship of Autonomy and Integrity in Medical Ethics." In: Allebeck P., Jansson B. (eds.). *Ethics in Medicine.* Raven Press: New York.
Peters T., 1997. *Playing God? Genetic Determinism and Human Freedom.* Routledge: New York.
Pokorski R.J., 1994. "Use of Genetic Information by Private Insurers." In: Murphy, Lappé (eds.), 1994.
President's Commission for the Study of Ethical Problems in Medicine and Biomedical and Behavioral Research, 1982. *Splicing Life: A Report on the Social and Ethical Issues of Genetic Engineeering with Human Beings.* U.S. Government Printing Office: Washington D.C.
Proctor R. N., 1991.*Value-Free Science? Purity and Power in Modern Knowledge.* Harvard University Press: Cambridge.
Rachels J., 1990. *Created from Animals: The Moral Implications of Darwinism.* Oxford University Press: Oxford.
Regan T, 1988 (2nd ed.). *The Case for Animal Rights.* Routledge: London.
Reich W.T. (ed.), 1995. *Encyclopedia of bioethics.* Macmillan: New York.
Reiss M.J., Straughan R., 1996. *Improving Nature? The science and ethics of genetic engineering.* Cambridge University Press: Cambridge.
Richter G., Bacchetta M.D., 1998. "Intervention in the Human Genome: Some Moral and Ethical Considerations." *Journal of Medicine and Philosophy* 23, 3, pp. 303-317.
Rifkin J., 1983. *Algeny.* Viking: New York.
Robertson J.A., 1994. *Children of Choice. Freedom and the New Reproductive Technologies.* Princeton University Press: Princeton.
Rollin B.E., 1995. *The Frankenstein syndrome: Ethical and social issues in the genetic engineering of*

animals. Cambridge University Press: New York.

Rolston III H., 1994. *Conserving Natural Value*. Columbia University Press: New York.

Rosch E., Lloyd B.B. (eds.), 1978. *Cognition and Categorization*. Lawrence Erlbaum: Hillsdale.

Rose S., Lewontin R.C., Kamin L.J., 1984. *Not in Our Genes: Biology, Ideology, and Human Nature*. Pelican, London.

Rosenberg A., 1996. "The Human Genome Project: Research Tactics and Economic Strategies." In: Frankel Paul, Miller, Paul (eds.), 1996.

Ross A. (ed.), 1996. *Science Wars*. duke University Press: Durham.

Rudolph F.B., McIntire L.V. (eds.), 1996. *Biotechnology: Science, Engineering, and Ethical Challenges for the 21st Century*. Joseph Henry Press: Washington D.C.

Russell W.M.S., Burch R.L., 1992 (2nd ed.) (1959). *The Principles of Humane Experimental Technique*. Methuen: London.

Russell B., 1960. "The Social Responsibilities of Scientists." *Science* 131, pp. 391-392.

Sagoff M., 1992. *The Economy of the Earth*. Cambridge University Press: Cambridge.

Scarr S., Weinberg R.A., 1976. "IQ Test Performance of Blach Children Adopted by White Families." *American Psychologist* 31, pp. 726-739.

Scarr S., 1992. "Development theories for the 1990s: Development and Individual Differences." *Child Development* 63, pp. 1-19.

Schweiker W., 1995. *Responsibility and Christian Ethics*. Cambridge University Press: Cambridge.

Segerstrahle U., 1994. "Scientific Intuition and Fraud." In: Erwin, Gendin, Kleiman (eds.), 1994.

Shiloh S., 1996. "Decision-making in the context of genetic risk." In: Marteau, Richards (eds.), 1996.

Shrader-Frechette K., 1994. *Ethics of Scientific Research*. Rowman & Littlefield Publ.: London.

Silver L.M., 1997. *Remaking Eden: Cloning and Beyond in a Brave New World*. Avon Books: New York.

Singer P., 1990 (2nd ed.). *Animal Liberation*. Cape: London.

Sinsheimer R.L., 1969. "The Prospect of Designed Genetic Change." In: Chadwick (ed.), 1994.

Smiley M., 1992. *Moral Responsibility and the Boundaries of Community: Power and Accountability from a Pragmatic Point of View*. The University of Chicago Press: Chicago.

Smith E., Sapp W. (eds.), 1997. *Plain Talk about the Human Genome Project: A Tuskegee University Conference on Its Promise and Perils... and Matters of Race*. Tuskegee University: Tuskegee.

Smith C.I.E., 1999. "Gene Therapy: A Molecular Perspective." In: Nordgren (ed.), 1999.

Sokal A., Bricmont J., 1998. *Intellectual Impostures: Postmodern philosophers' abuse of science*. Profile Books: London.

Sontag S., 1990. *Illness as Metaphor; and, AIDS and Its Metaphors*. Doubleday Anchor Books: New York.

Steen R.G., 1996. *DNA and Destiny: Nature and Nurture in Human Behavior*. Plenum Press: New York.

Stevenson L., Byerly H., 1995. *The Many Faces of Science: An Introduction to Scientists, Values, and Society*. Westview Press: Boulder.

Stone D.A., 1996. "The Implications of the Human Genome Project for Access to Health Insurance." In: Murray, Rothstein, Murray (eds.), 1996.

Strohman R.C., 1993. "Ancient Genomes, Wise Bodies, Unhealthy People: Limits of a Genetic Paradigm in Biology and Medicine." *Perspectives in Biology and Medicine* 37, 1, pp. 112-145.

Strong C., 1997. *Ethics in Reproductive and Perinatal Medicine: A New Framework*. Yale University Press: New Haven.

Svatos M., 1996. "Biotechnology and the Utilitarian Argument for Patents." In: Frankel Paul, Miller, Paul (eds.), 1996.

Taylor P.W., 1986. *Respect for Nature: A Theory of Environmental Ethics*. Princeton University Press: Princeton.

Thom D., Jennings M., 1996. "Human pedigree and the 'best stock': from eugenics to genetics." In: Marteau, Richards (eds.), 1996.

Thomasma D.C., Kushner T., 1996. *Birth to Death: Science and Bioethics*. Cambridge University Press: Cambridge.

Tong R., 1997. *Feminist Approaches to Bioethics: Theoretical Reflections and Practical Applications*. Westview Press: Boulder.

Toon P.D., 1993. "After bioethics and towards virtue?" *Journal of Medical Ethics* 19, pp. 17-18.

Torres J.M., 1997. "On the Limits of Enhancement in Human Gene Transfer: Drawing the Line." *The Journal of Medicine and Philosophy* 22, pp. 43-53.

Varela F.J., 1999. *Ethical Know-How: Action, Wisdom, and Cognition*. Stanford University Press: Stanford.

Venter J.C. et al., 2001. "The Human Genome: The Sequence of the Human Genome." *Science* Vol. 291 No. 5507 (16 February), pp. 1304-1351.

Wachbroit R., 1993. "Rethinking Medical Confidentiality: The Impact of Genetics." *Suffolk University Law Review* XXVII, pp. 1391-1410.

Wachbroit R., 1994a. "Normality as a Biological Concept." *Philosophy of Science* 61, 4, pp. 579-591.

Wachbroit R.S., 1994b. "Distinguishing Genetic Disease and Genetic Susceptibility." *American Journal of Medical Genetics* 53, pp. 236-240.

Walters L., 1986. "The Ethics of Human Gene Therapy." *Nature* 320, pp. 225-227.

Walters L., 1999. "Human Gene Therapy: Past, Present, and Future." In: Nordgren (ed.), 1999.

Walters L., Palmer J.G., 1997. *The Ethics of Human Gene Therapy*. Oxford University Press: Oxford.

Warren M.A., 1982. "Future Generations". In: Regan T., VanDeVeer D. (eds.), 1982. *And Justice For All*. Rowman and Allanheld: Totowa, NJ.

Weiner C., 1994. "Anticipating the Consequences of Genetic Engineering. Past, Present, and Future." In: Cranor, C. (ed.). *Are Genes Us? The Social Consequences of the New Genetics*. Rutgers University Press: New Brunswick.

Weiner C., 1999. "Social Responsibility in Genetic Engineering: Historical Perspectives." In: Nordgren (ed.), 1999.

Weir R.F., Lawrence S.C., Fales E. (eds.), 1994. *Genes and Human Self-Knowledge: Historical and Philosophical Reflections on Modern Genetics*. University of Iowa Press: Iowa City.

Welin S. (ed.), 1993. *Scientific Responsibility and Public Control*. Studies in Research Ethics No 2. Centre for Research Ethics: Göteborg.

Werhane P., Doering J., 1997. "Conflicts of Interest and Conflicts of Commitment." In Elliott, Stern (eds.), 1997.

Wildes K.W., 1993. "The Priesthood of Bioethics and the Return of Casuistry." *The Journal of Medicine and Philosophy* 18, pp. 39-40.'

Wilkinson T., 1998. *Science Under Siege: The Politicians' War on Nature and Truth*. Johnson Books: Boulder.

Willigenburg T. van, Heeger R., 1991. "Rechtfertigung moralischer Urteile. Ein Netzmodell." *Zeitschrift für Evangelische Ethik* 35, pp. 88-95.

Wilson E.O., 1978. *On Human Nature*. Harvard University Press: Cambridge.

Winkler E.R., 1993. "From Kantianism to Contextualism: The Rise and Fall of the Paradigm theory in Bioethics." In: Winkler E.R., Coombs J.R. (eds.), 1993. *Applied Ethics: A Reader*. Blackwell: Oxford.

Wittgenstein L., 1953. *Philosophical Investigations*. Macmillan: New York.

World Commission on Environment and Development, 1987. *Our Common Future*. Oxford University Press: Oxford.

Wright R., 1990. "Achilles' Helix." *The New Republic*, 9 and 16 July, 21-31.

Zilinskas R.A., 1997. "Global Perspectives on the Human Genome Project." In: Smith, Sapp (eds.), 1997.

Zutphen L.F.M. van, Meer M. van der, 1997. *Welfare Aspects of Transgenic Animals: Proceedings EC Workshop of October 30, 1995*. Springer-Verlag: Berlin.

INDEX

Philosophy and Medicine

39. M.A.G. Cutter and E.E. Shelp (eds.): *Competency*. A Study of Informal Competency Determinations in Primary Care. 1991 ISBN 0-7923-1304-6
40. J.L. Peset and D. Gracia (eds.): *The Ethics of Diagnosis*. 1992 ISBN 0-7923-1544-8
41. K.W. Wildes, S.J., F. Abel, S.J. and J.C. Harvey (eds.): *Birth, Suffering, and Death*. Catholic Perspectives at the Edges of Life. 1992 [CSiB-1] ISBN 0-7923-1547-2; Pb 0-7923-2545-1
42. S.K. Toombs: *The Meaning of Illness*. A Phenomenological Account of the Different Perspectives of Physician and Patient. 1992 ISBN 0-7923-1570-7; Pb 0-7923-2443-9
43. D. Leder (ed.): *The Body in Medical Thought and Practice*. 1992 ISBN 0-7923-1657-6
44. C. Delkeskamp-Hayes and M.A.G. Cutter (eds.): *Science, Technology, and the Art of Medicine*. European-American Dialogues. 1993 ISBN 0-7923-1869-2
45. R. Baker, D. Porter and R. Porter (eds.): *The Codification of Medical Morality*. Historical and Philosophical Studies of the Formalization of Western Medical Morality in the 18th and 19th Centuries, Volume One: Medical Ethics and Etiquette in the 18th Century. 1993 ISBN 0-7923-1921-4
46. K. Bayertz (ed.): *The Concept of Moral Consensus*. The Case of Technological Interventions in Human Reproduction. 1994 ISBN 0-7923-2615-6
47. L. Nordenfelt (ed.): *Concepts and Measurement of Quality of Life in Health Care*. 1994 [ESiP-1] ISBN 0-7923-2824-8
48. R. Baker and M.A. Strosberg (eds.) with the assistance of J. Bynum: *Legislating Medical Ethics*. A Study of the New York State Do-Not-Resuscitate Law. 1995 ISBN 0-7923-2995-3
49. R. Baker (ed.): *The Codification of Medical Morality*. Historical and Philosophical Studies of the Formalization of Western Morality in the 18th and 19th Centuries, Volume Two: Anglo-American Medical Ethics and Medical Jurisprudence in the 19th Century. 1995 ISBN 0-7923-3528-7; Pb 0-7923-3529-5
50. R.A. Carson and C.R. Burns (eds.): *Philosophy of Medicine and Bioethics*. A Twenty-Year Retrospective and Critical Appraisal. 1997 ISBN 0-7923-3545-7
51. K.W. Wildes, S.J. (ed.): *Critical Choices and Critical Care*. Catholic Perspectives on Allocating Resources in Intensive Care Medicine. 1995 [CSiB-2] ISBN 0-7923-3382-9
52. K. Bayertz (ed.): *Sanctity of Life and Human Dignity*. 1996 ISBN 0-7923-3739-5
53. Kevin Wm. Wildes, S.J. (ed.): *Infertility: A Crossroad of Faith, Medicine, and Technology*. 1996 ISBN 0-7923-4061-2
54. Kazumasa Hoshino (ed.): *Japanese and Western Bioethics*. Studies in Moral Diversity. 1996 ISBN 0-7923-4112-0

Philosophy and Medicine

55. E. Agius and S. Busuttil (eds.): *Germ-Line Intervention and our Responsibilities to Future Generations.* 1998 ISBN 0-7923-4828-1
56. L.B. McCullough: *John Gregory and the Invention of Professional Medical Ethics and the Professional Medical Ethics and the Profession of Medicine.* 1998 ISBN 0-7923-4917-2
57. L.B. McCullough: *John Gregory's Writing on Medical Ethics and Philosophy of Medicine.* 1998 [CiME-1] ISBN 0-7923-5000-6
58. H.A.M.J. ten Have and H.-M. Sass (eds.): *Consensus Formation in Healthcare Ethics.* 1998 [ESiP-2] ISBN 0-7923-4944-X
59. H.A.M.J. ten Have and J.V.M. Welie (eds.): *Ownership of the Human Body.* Philosophical Considerations on the Use of the Human Body and its Parts in Healthcare. 1998 [ESiP-3] ISBN 0-7923-5150-9
60. M.J. Cherry (ed.): *Persons and Their Bodies.* Rights, Responsibilities, Relationships. 1999 ISBN 0-7923-5701-9
61. R. Fan (ed.): *Confucian Bioethics.* 1999 [APSiB-1] ISBN 0-7923-5853-8
62. L.M. Kopelman (ed.): *Building Bioethics.* Conversations with Clouser and Friends on Medical Ethics. 1999 ISBN 0-7923-5853-8
63. W.E. Stempsey: *Disease and Diagnosis.* 2000 PB ISBN 0-7923-6322-1
64. H.T. Engelhardt (ed.): *The Philosophy of Medicine.* Framing the Field. 2000 ISBN 0-7923-6223-3
65. S. Wear, J.J. Bono, G. Logue and A. McEvoy (eds.): *Ethical Issues in Health Care on the Frontiers of the Twenty-First Century.* 2000 ISBN 0-7923-6277-2
66. M. Potts, P.A. Byrne and R.G. Nilges (eds.): *Beyond Brain Death.* The Case Against Brain Based Criteria for Human Death. 2000 ISBN 0-7923-6578-X
67. L.M. Kopelman and K.A. De Ville (eds.): *Physician-Assisted Suicide.* What are the Issues? 2001 ISBN 0-7923-7142-9
68. S.K. Toombs (ed.): *Handbook of Phenomenology and Medicine.* 2001 ISBN 1-4020-0151-7; Pb 1-4020-0200-9
69. R. ter Meulen, W. Arts and R. Muffels (eds.): *Solidarity in Health and Social Care in Europe.* 2001 ISBN 1-4020-0164-9
70. A. Nordgren: *Responsible Genetics.* The Moral Responsibility of Geneticists for the Consequences of Human Genetics Research. 2001 ISBN 1-4020-0201-7

KLUWER ACADEMIC PUBLISHERS – DORDRECHT / BOSTON / LONDON

MIX
Papier aus verantwortungsvollen Quellen
Paper from responsible sources
FSC® C105338

If you have any concerns about our products,
you can contact us on
ProductSafety@springernature.com

In case Publisher is established outside the EU,
the EU authorized representative is:
Springer Nature Customer Service Center GmbH
Europaplatz 3, 69115 Heidelberg, Germany

Printed by Libri Plureos GmbH
in Hamburg, Germany